About City & Guilds

City & Guilds is the UK's leading provider of vocational qualifications, offering over 500 awards across a wide range of industries, and progressing from entry level to the highest levels of professional achievement. With over 8500 centres in 100 countries, City & Guilds is recognised by employers worldwide for providing qualifications that offer proof of the skills they need to get the job done.

Equal opportunities

City & Guilds fully supports the principle of equal opportunities and we are committed to satisfying this principle in all our activities and published material. A copy of our equal opportunities policy statement is available on the City & Guilds website.

Copyright

First edition 2014

ISBN 978-0-85193-278-1

Publisher: Charlie Evans

Development Editor: Hannah Cooper

Production Editor: Fiona Freel

Picture Research: Katherine Hodges

Project Management and Editorial Series Team: Vicky Butt, Anna Clark, Kay Coleman, Jo Kemp, Karen Hemingway, Jon Ingoldby, Caroline Low, Joan Miller, Shirley Wakley

Cover design by Select Typesetters Ltd

Text design by Design Deluxe, Bath

Indexed by Indexing Specialists (UK) Ltd

Illustrations by Saxon Graphics Ltd and Ann Paganuzzi

Typeset by Saxon Graphics Ltd, Derby

Printed in the UK by Cambrian Printers Ltd

Publications

For information about or to order City & Guilds support materials, contact 0844 534 0000 or centresupport@cityandguilds.com. You can find more information about the materials we have available at www.cityandguilds.com/publications.

Every effort has been made to ensure that the information contained in this publication is true and correct at the time of going to press. However, City & Guilds' products and services are subject to continuous development and improvement and the right is reserved to change products and services from time to time. City & Guilds cannot accept liability for loss or damage arising from the use of information in this publication.

City & Guilds
1 Giltspur Street
London EC1A 9DD

T 0844 543 0033
www.cityandguilds.com
publishingfeedback@cityandguilds.com

THE CITY & GUILDS TEXTBOOK

LEVEL 3 NVQ DIPLOMA IN
ELECTROTECHNICAL TECHNOLOGY 2357
UNITS 301–304

PAUL HARRIS, ANDREW HAY-ELLIS AND TREVOR PICKARD
SERIES EDITOR: PETER TANNER

ACKNOWLEDGEMENTS

City & Guilds would like to thank sincerely the following:

For invaluable knowledge and expertise

Gary Willard, *JP, Dip.EE, CEng, MIEE*, The Institution of Engineering and Technology, Technical Reviewer

Richard Woodcock, City & Guilds, Technical Reviewer and Contributor

For supplying pictures for the front and back covers

Jules Selmes

For their help with photoshoots

Jules Selmes and Adam Giles (photographer and assistant);

Andy Jeffery, Ben King and students from Oaklands College, St Albans; Jaylec Electrical; Andrew Hay-Ellis, James L Deans and the staff at Trade Skills 4 U, Crawley, and the following models: Jordan Hay-Ellis, Terry White, Katherine Hodges, Claire Owen.

Picture credits

Every effort has been made to acknowledge all copyright holders as below and the publishers will, if notified, correct any errors in future editions.

BSI (Permission to reproduce extracts from British Standards is granted by BSI Standards Limited (BSI). No other use of this material is permitted. British Standards can be obtained in PDF or hard copy formats from the BSI online shop: www.bsigroup.com/Shop or by contacting BSI Customer Services for hard copies only: Tel: +44 (0)20 8996 9001, Email: cservices@bsigroup.com); **BSRIA** p164, p165; **Ecoplay Nederland B.V.** p166; **HSE** p219; **IET** p7, p27, p219, p228, p229, p302, p303, p304, p321, p322, p323; **Insulated Tools Ltd** p16; **Kensa Engineering Ltd** p113; **Lasnek** p348, p350; **Marshall-Tufflex** p345, p347, p350, p351; **Prysimian** p343; **ScrewFix** p12, p38, p39; **Shutterstock** p1, p12, p46, p79, p88, p92, p95, p108, p110, p116, p128, p130, p131, p132, p133, p144, p151, p177, p201, p204, p217, p227, p253, p345, p346, p351; **Worcester Bosch Thermotechnology Ltd** p110; **UK Power Networks** p194.

Text permissions

For kind permission of text extracts:

Crown copyright. Contains public sector information published by the Health and Safety Executive and licensed under the Open Government Licence v.1.0 on page 24 and includes extracts from the following HSE publications: 'Construction information sheet no. 59', p197; 'Electricity Safety, Quality and Continuity Regulations' p258, 259.

Crown copyright: 'Building Regulations' p98.

Permission to reproduce extracts from British Standards is granted by BSI Standards Limited (BSI). No other use of this material is permitted. British Standards can be obtained in PDF or hard copy formats from the BSI online shop: www.bsigroup.com/Shop or by contacting BSI Customer Services for hard copies only: Tel: +44 (0) 845 086 9001, Email: cservices@bsigroup.com. Pages 180, 193, 277, 309, 316, 318, 327, 330, 331.

BS 7671: 2008 Incorporating Amendment No 1: 2011 can be purchased in hardcopy format only from the IET website http://electrical.theiet.org/ and the BSI online shop: http://shop.bsigroup.com

From the authors

Paul Harris: I would like to thank my wife, Carol, and sons, George and Alfie, for their support and for putting up with me while writing some of the sections.

Andrew Hay-Ellis: Many thanks to my wife, Michelle, for her patience and understanding during the many hours spent writing; also to my sons, Jordan and Brendan, for proofreading sections of text to make sure the words said what I intended.

Trevor Pickard: I need to express my sincere thanks to my wife, Sue, whose command of the English language has always been better than mine and without whose help the job of the copyeditor would have been that much harder!

Peter Tanner: Many thanks to my wife, Gillian, and daughters, Rebecca and Lucy, for their incredible patience; also to Jim Brooker, my brother-in-law, for showing me the importance of carrying out the correct safe isolation procedure on more than one occasion!

CONTENTS

Contents

ABOUT THE AUTHORS

Eur Ing **Paul Harris** *BEng (Hons), CEng, FIHEEM, MIEE, MCIBSE*

I am a Chartered Electrical Engineer, having started my career as a JIB Indentured Apprentice in 1978.

Following completion of my apprenticeship I worked in a variety of electrical installation jobs. Throughout my apprenticeship and time as a craftsman, through my enthusiasm to learn, I have always questioned 'why?', 'how?' and 'what will happen if…?' I believe it is important not to separate theory and learning from work life, and to take pride as an individual in work, development and the profession.

I returned to college in 1989, followed by university, and obtained an HNC Electrical Engineering and later BEng (Hons) in Building Services Engineering by part-time study.

In my work life I have held a number of engineer posts including managing direct labour work forces and teams of professional engineers. I have been a part-time lecturer for CGLI and BTEC ONC electrical Installation and have also written technical articles for a variety of institutions.

I currently have my own consultancy where I provide design expertise in specialist areas and installations to a variety of clients and industry professionals. Alongside this I am a Medical Locations expert for Working Group 710, represent the UK at IEC and am a committee member of JPEL64, which is responsible for the production of BS 7671.

My belief held in 1978 remains unchanged – training and development, with its supporting books and documentation, is invaluable to individuals and employers alike and should be valued throughout your career.

Andrew Hay-Ellis *CertEd, QTLS, MIET, LCGI, MinstLM*

When I first left school I joined a firm of chartered accountants (an excuse to play cricket) and after three years I moved to a small mechanical and electrical company primarily to look after the company's day-to-day accounts. As with many small companies, staff members often end up working wherever there is a need and so I soon found myself out on site with the engineers. Looking back to my childhood I always had an enquiring mind and electrical installation work provided the mental and physical challenges that I needed and was no longer finding in accountancy. Some thirty plus years later, having worked as an electrician, a supervisor and a project engineer on both commercial and industrial installations, I can honestly say this was the best decision I could have made.

An insatiable thirst for knowledge and new challenges led me back to college to improve my qualifications and this led to my next career change, moving from electrical contracting into teaching. I have now been teaching for over seventeen years, working both in private training and in further education. While my current roles as Director of Education for Trade Skills 4U and within City & Guilds are more technical, I still very much enjoy the thrill of teaching.

The job of an electrician is becoming ever more technical and the need for quality training is a must for anyone working within the electrical industry. Open your mind to the possibilities and never stop asking why. Set your sights high and strive to achieve and you will never be disappointed with the choices you have made.

Trevor Pickard *IEng, MIET*

I am an electrical engineering consultant and my interest in all things electrical started when I was quite young. I always had a battery powered model under construction or an electrical motor or some piece of electrical equipment in various stages of being taken apart to see how they operated. Looking back, some of my activities with mains electricity would certainly be considered as unacceptable today!

Upon leaving school in 1966 I commenced work with an electricity distribution company, Midlands Electricity Board (MEB) and after serving a student apprenticeship I held a series of engineering positions. I have never tired of my involvement with electrical engineering and was very fortunate to have had a varied and interesting career in the engineering department of Midlands Electricity and embraced its various changes through privatisation and subsequent acquisition. I held posts in Design, Safety, Production Engineering, as Production Manager of a large urban-based operational division, as General Manager of the Repair and Restoration department, and as General Manager of the Primary Network department (33kV–132kV).

My interest in electrical engineering has extended beyond the '9–5 job' and I have had the opportunity to become involved in the writing of standards in the domestic, European and international arena with BSI, CENELEC and IEC and have for many years lectured for the Institution of Engineering and Technology.

Electricity is with us in almost every aspect of modern life and for those who are just starting their career in this field I would say keep an open mind, be safety conscious in how you carry out your work and who knows where your studies will take you.

Peter Tanner *MIET, LCGI*

Series Editor

I started in the industry while still at school, chasing walls for my brother-in-law for a bit of pocket-money. This taught me quickly that if I took a career in the industry I needed to progress as fast as I could.

Jobs in the industry were few and far between when I left school so after a spell in the armed forces, I gained a place as a sponsored trainee on the CITB training scheme. I attended block release at Guildford Technical College where the CITB would find me work experience with 'local' employers. My first and only work experience placement was with a computer installation company located over twenty miles away so I had to cycle there every morning but I was desperate to learn and enjoyed my work.

Computer installations were very different in those days. Computers filled large rooms and needed massive armoured supply cables so the range of work I experienced was vast from data cabling, to all types of containment systems and low voltage systems.

In the second year of my apprenticeship I found employment with a company where most of my work centred around the London area. The work was varied, from lift systems in well-known high-rise buildings to lightning protection on the side of even higher ones!

On completion of my apprenticeship I worked for a short time as an intruder alarm installer mainly in domestic dwellings, a role where client relationships and handling information is very important.

Following this I began work with a company where I was involved in shop-fitting and restaurant and pub refurbishments. It wasn't long before I was managing jobs and gaining further qualifications through professional development. I was later seconded to the Property Services Agency designing major installations within some of the most well-known buildings in the UK.

A career-changing accident took me into teaching where I truly found the rewards the industry has to offer. Seeing young trainees maturing into qualified electricians is a worthwhile experience. On many occasions I see many of my old trainees when they attend further training and update courses. Seeing their successes makes it all worthwhile.

I have worked with City & Guilds for over twenty years and represent them on a variety of industry committees such as JPEL64, which is responsible for the production of BS 7671. I am passionate about using my vast experience in the industry to maintain the high standards the industry expects.

HOW TO USE THIS TEXTBOOK

Welcome to your City & Guilds Level 3 NVQ Diploma in Electrotechnical Technology textbook. It is designed to guide you through your 2357 Level 3 qualification and be a useful reference for you throughout your career.

Each chapter covers a unit from the 2357 Level 3 qualification. Each chapter covers everything you will need to understand in order to complete your written or online tests and prepare for your practical assessments. Across some learning outcomes in the 2357 units there is some natural revisiting of knowledge and skills in different contexts, which the content in this book also reflects, for practical use and reference.

Throughout this textbook you will see the following features:

KEY POINT

Micro-CHP boilers are only suitable where there is a high demand for heat.

KEY POINT These are particularly useful points that may assist in you in revision for your tests or to help you remember something important.

Bill of quantity

A contract document comprising a list of materials required for the works and their estimated quantities.

DEFINITIONS Words in bold in the text are explained in the margin to aid your understanding. They also appear in the glossary at the back of the book.

ACTIVITY

Which parts of the Building Regulations apply to electrical installations?

ACTIVITY These provide questions or suggested activities to help you learn and practise.

ASSESSMENT GUIDANCE

You must be able to distinguish between statutory and non-statutory regulations.

ASSESSMENT GUIDANCE These highlight useful points that are helpful for your learning and assessment.

 Smartscreen Unit 301
Handout 1

SMARTSCREEN These provide references to City & Guilds online learner and tutor resources, which you can access on SmartScreen.co.uk.

Assessment criteria

3.8 Describe safe practices and procedures in the working environment.

ASSESSMENT CRITERIA These highlight the assessment criteria coverage through each unit, so you can easily link your learning to what you need to know or do for each Learning outcome.

You will also see the following abbreviation in the running heads:

LO – learning outcome   **LO4**   **Procedures for a safe working environment**

Where tables and forms in this book have been used directly from other publications such as the IET this has been noted, and the style reflects the original (with kind permission). Always make sure you use the latest information and forms.

UNIT 301
Understanding health and safety legislation, practices and procedures

Every year, accidents at work result in the deaths of more than one hundred people, with several hundred thousand more being injured in the workplace. In 2010/2011, Health and Safety Executive (HSE) statistics recorded 27 million working days being lost due to work-related illness and workplace injury.

Occupational health and safety affects all individuals in the workplace and all aspects of work in the complete range of working environments – hospitals, factories, schools, universities, commercial undertakings, manufacturing plants and offices. As well as the human cost in terms of pain, grief and suffering, accidents in the workplace also have a financial cost, such as lost income, insurance and production disturbance. The HSE put this figure at £13.4 billion for the year 2010/2011. It is therefore important to identify, assess and control the activities that may cause harm in the workplace.

This unit is designed to enable you to understand health and safety legislation and associated practices and procedures, when installing and maintaining electrotechnical systems and equipment. You will need this knowledge to underpin the application of health and safety legislation, practices and procedures.

LEARNING OUTCOMES

There are four learning outcomes to this unit. The learner will understand:

1 how relevant health and safety legislation applies in the workplace
2 the procedures for dealing with health and safety in the work environment
3 the procedures for establishing a safe working environment
4 the requirements for identifying and dealing with hazards in the work environment.

This unit will be assessed by:

■ an assignment
■ online multiple-choice assessment.

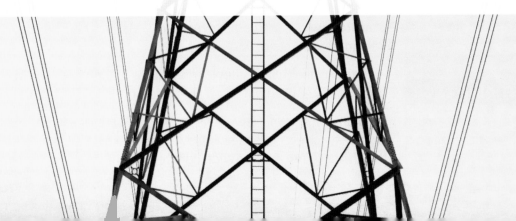

Know how relevant health and safety legislation applies in the workplace

The basic concept of health and safety legislation is to provide the legal framework for the protection of people from illness and physical injury that may occur in the workplace.

ROLES AND RESPONSIBILITIES WITH REGARD TO CURRENT RELEVANT LEGISLATION

There are two sub-divisions of the law that apply to health and safety: civil law and criminal law.

Civil law – deals with disputes between individuals, between organisations, or between individuals and organisations, in which compensation can be awarded to the victim. The civil court is concerned with **liability** and the extent of that liability rather than guilt or non-guilt.

Liability

A debt or other legal obligation in order to compensate for harm.

Criminal law – is the body of rules that regulates social behaviour and prohibits threatening, harming or other actions that may endanger the health, safety and moral welfare of people. The rules are laid down by the government and are enacted by Acts of Parliament as **statutes**. The Health and Safety at Work etc Act 1974 (HSW Act) is an example of criminal law. It is enforced by the Health and Safety Executive (HSE) or Local Authority environmental health officers.

There are two sources of law for both the civil and criminal law areas. These are common law and statute law.

Statute

A major written law passed by Parliament.

Common law – is the body of law based on custom and decisions made by judges in courts. In health and safety, the legal definitions of terms such as 'negligence', 'duty of care', and 'so as far as is reasonably practicable' are based on legal judgments and are part of common law.

KEY POINT

'So far as is reasonably practicable' involves weighing a risk against the trouble, time and money needed to control it.

Statute law – is the name given to law that has been laid down by Parliament as Acts of Parliament.

In terms of health and safety, criminal law is based only on statute law, but civil law may be based on either common law or statute law.

In summary, criminal law seeks to protect everyone in society and civil law seeks to recompense individuals, to make amends for loss or harm they have suffered (ie provide compensation).

ASSESSMENT GUIDANCE

Remember that everyone on site is responsible for their and everyone else's safety.

The main legal requirements for health and safety at work

The HSW Act is the basis of all British health and safety law. It provides a comprehensive and integrated piece of legislation that sets out the general duties that employers have towards employees, contractors and members of the public, and that employees have towards themselves and each other. These duties are qualified in the HSW Act by the principle of 'so far as is reasonably practicable'.

What the law expects is what good management and common sense would lead employers to do anyway, that is, to look at what the risks are and take sensible measures to tackle those risks. The person(s) who is responsible for the risk and best placed to control that risk is usually designated as the **duty holder**.

The HSW Act lays down the general legal framework for health and safety in the workplace, with specific duties being contained in regulations, also called statutory instruments (SIs), which are also examples of laws approved by Parliament.

Duty holder

The person in control of the danger is the duty holder. This person must be competent by formal training and experience and with sufficient knowledge to avoid danger. The level of competence will differ for different items of work.

 SmartScreen Unit 301
Handout 1

INDIVIDUALS' RESPONSIBILITIES UNDER HEALTH AND SAFETY LEGISLATION

The HSW Act, which is an **enabling Act**, is based on the principle that those who create risks to employees or others in the course of carrying out work activities are responsible for controlling those risks. The HSW Act places specific responsibilities on:

- employers
- the self-employed
- employees
- designers
- manufacturers and suppliers
- importers.

This section will deal with the responsibilities of employers, the self-employed and employees.

Enabling Act

An enabling Act allows the Secretary of State to make further laws (regulations) without the need to pass another Act of Parliament.

The HSW Act makes provision for securing the health, safety and welfare of persons at work

Responsibilities of employers and the self-employed

Under the main provisions of the HSW Act, employers and the self-employed have legal responsibilities in respect of the health and safety of their employees and other people (eg visitors and contractors) who may be affected by their undertaking and exposed to risks as a result. The employers' general duties are contained in Section 2 of the Act.

They are to ensure, 'so far as is reasonably practicable', the health, safety and welfare at work of all their employees, with particular regard to:

- the provision of safe plant and systems of work
- the safe use, handling, storage and transport of articles and substances
- the provision of any required information, instruction, training or supervision
- a safe place of work including safe access and exit
- a safe working environment with adequate welfare facilities.

These duties apply to virtually everything in the workplace, which therefore includes electrical systems and installations, plant and equipment. An employer does not have to take measures to avoid or reduce the risk if that is technically impossible or if the time, trouble or cost of the measures would be grossly disproportionate to the risk.

Responsibilities of employees

Employees are required to take reasonable care for the health and safety of themselves and others. To achieve this aim, they have two main duties placed upon them:

- to take reasonable care for the health and safety of themselves and others who may be affected by their acts or omissions at work
- to cooperate with their employer and others to enable them to fulfil their legal obligations.

In addition there is a duty not to misuse or interfere with safety provisions.

Most of the duties in the HSW Act and the general duties included in the Management of Health and Safety at Work Regulations 1999 (the Management Regulations) are expressed as goals or targets that are to be met 'so far as is reasonably practicable' or through exercising 'adequate control' or taking 'appropriate' (or 'reasonable') steps.

This involves making judgments as to whether existing control measures are sufficient and, if not, deciding what else should be done to eliminate or reduce the risk.

This risk assessment will be produced using approved codes of practice (ACoP) and published standards, as well as HSE or industry guidance on good practice, where available.

STATUTORY AND NON-STATUTORY HEALTH AND SAFETY MATERIALS

When the HSW Act came into force there were already some 30 statutes and 500 sets of regulations in place. The aim of the Health and Safety Commission (HSC) and the Health and Safety Executive (HSE) was to progressively replace the existing regulations with a system of regulation that expresses general duties, principles and goals, with any supporting detail set out in ACoPs and guidance.

Regulations

Statutory instruments (SIs) are laws approved by Parliament. The regulations governing health and safety are usually made under the HSW Act, following proposals from the HSC/HSE. This applies to regulations based on European Commission (EC) Directives as well as those produced in Great Britain.

The HSW Act, and general duties in the Management Regulations, set goals and leave employers the freedom to decide how to control the risks they identify. Guidance and ACoPs give advice.

Some risks are considered so great or the proper control measures so costly that it would not be appropriate to leave employers to decide what to do about them. Regulations identify these risks and set out the specific action that must be taken. Often these requirements are absolute – they require something to be done, without qualification. The employer has no choice but to undertake whatever action is required to prevent injury, regardless of cost or effort.

Some regulations apply across all workplaces. Such regulations include the Manual Handling Operations Regulations 1992, which apply wherever things are moved by hand or bodily force, and the Health and Safety (Display Screen Equipment) Regulations 1992, which apply wherever visual display units (VDUs) are used. Other regulations apply to hazards unique to specific industries, such as mining or the nuclear industry.

The following regulations apply across the full range of workplaces.

- **Control of Noise at Work Regulations 2005**: require employers to take action to protect employees from hearing damage.

- **Control of Substances Hazardous to Health (COSHH) Regulations 2002 (as amended)**: require employers to assess the risks from hazardous substances and take appropriate precautions.

- **Electricity at Work Regulations 1989**: require people in control of electrical systems to ensure they are safe to use and maintained in a safe condition.

Using personal protective equipment (PPE) while working

- **Health and Safety (Display Screen Equipment) Regulations 1992**: gives specific requirements for the use of display equipment such as computer screens. This may effect the choice of lighting to reduce glare.

- **Health and Safety (First-Aid) Regulations 1981**: require employers to provide adequate and appropriate equipment, facilities and personnel to ensure their employees receive immediate attention if they are injured or taken ill at work. These regulations apply to all workplaces, including those with fewer than five employees, and to the self-employed.
- **Health and Safety Information for Employees Regulations 1989**: require employers to display a poster telling employees what they need to know about health and safety.
- **Management of Health and Safety at Work Regulations 1999 (as amended)**: require employers to carry out risk assessments, make arrangements to implement necessary measures, appoint competent people and arrange for appropriate information and training.
- **Manual Handling Operations Regulations 1992**: cover the moving of objects by hand or bodily force.
- **Personal Protective Equipment at Work Regulations 1992 (as amended)**: require employers to provide appropriate protective clothing and equipment for their employees.

- **Provision and Use of Work Equipment Regulations 1998**: require that equipment provided for use at work, including machinery, is suitable and safe.
- **Reporting of Injuries, Diseases and Dangerous Occurrences Regulations 2013**: require employers to notify the HSE of certain occupational injuries, diseases and dangerous events.
- **Workplace (Health, Safety and Welfare) Regulations 1992**: cover a wide range of basic health, safety and welfare issues such as ventilation, heating, lighting, workstations, seating and welfare facilities.

The following specific regulations cover particular areas, such as asbestos and lead:

- **Chemicals (Hazard Information and Packaging for Supply) Regulations 2002**: require suppliers to classify, label and package dangerous chemicals and provide safety data sheets for them.
- **Construction (Design and Management) Regulations 2007**: cover safe systems of work on construction sites.
- **Control of Asbestos Regulations 2012**: affect anyone who owns, occupies, manages or otherwise has responsibilities for the maintenance and repair of buildings that may contain asbestos.
- **Control of Lead at Work Regulations 2002**: imposes duties on employers to carry out risk assessments, prevent or control exposure to lead and monitor the exposure of employees.

- **Control of Major Accident Hazards Regulations 1999 (as amended)**: require those who manufacture, store or transport dangerous chemicals or explosives in certain quantities to notify the relevant authority.
- **Dangerous Substances and Explosive Atmospheres Regulations 2002**: require employers and the self-employed to carry out a risk assessment of work activities involving dangerous substances.
- **Gas Safety (Installation and Use) Regulations 1998**: cover safe installation, maintenance and use of gas systems and appliances in domestic and commercial premises.
- **Work at Height Regulations 2005**: apply to all work at height where there is a risk of a fall liable to cause personal injury.

Approved codes of practice (ACoP)

ACoPs offer practical examples of good practice. They were made under Section 16 of the HSW Act and have a special status. They give advice on how to comply with the law by, for example, providing a guide to what is reasonably practicable. For example, if regulations use words such as 'suitable' and 'sufficient', an ACoP can illustrate what is required in particular circumstances. If an employer is prosecuted for a breach of health and safety law, and it is proved that they have not followed the provisions of the relevant ACoP, a court can find them at fault unless they can show that they have complied with the law in some other way.

Guidance and non-statutory regulations

The HSE publishes guidance on a range of subjects. Guidance can be specific to the health and safety problems of an industry or to a particular process used in a number of industries.

The main purposes of guidance are:

- to interpret and help people to understand what the law says
- to help people comply with the law
- to give technical advice.

Following guidance is not compulsory and employers are free to take other action, but if they do follow the guidance, they will normally be doing enough to comply with the law.

One very good example of guidance and non-statutory regulation is BS 7671 Requirements for Electrical Installations (the IET Wiring Regulations 17th edition). If electrotechnical work is undertaken in accordance with BS 7671, it is likely to meet the requirements of the Electricity at Work Regulations 1989, which deal with work with electrical equipment and systems.

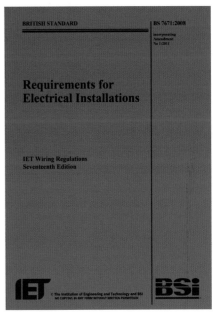

The IET Wiring Regulations

BS 7671 is the national standard in the UK for low voltage electrical installations. The document is largely based on documents produced by the European Committee for Electrotechnical Standardisation (CENELEC). The regulations deal with the design, selection, erection, inspection and testing of electrical installations operating at a voltage up to 1000 V a.c.

European law

In recent years much of Great Britain's health and safety law has originated in Europe. Proposals from the EC may be agreed by member states, which are then responsible for making them part of their domestic law.

Modern health and safety law in this country, including much of that from Europe, is based on the principle of risk assessment as required by the Management of Health and Safety at Work Regulations 1999.

ROLES OF THE HSE IN ENFORCING HEALTH AND SAFETY LEGISLATION

The Health and Safety Executive (HSE) and the Health and Safety Commission (HSC) were, until 2008, the two government agencies responsible for health and safety in Great Britain. The non-executive HSC was there to ensure that relevant legislation was appropriate and understood by conducting and sponsoring research, providing training, providing an information and advisory service, and submitting proposals for new or revised regulations (statutory instruments) and approved codes of practice (ACoPs).

The HSE was the operating arm of the HSC. It advised and assisted the HSC in its functions and had specific responsibility, shared with local authorities, for enforcing health and safety law. The HSC and HSE merged on 1 April 2008 and are now known simply as HSE.

Today, the HSE's aim is to prevent death, injury and ill health in Great Britain's workplaces and it has a number of ways of achieving this. Enforcing authorities may offer the duty holder information and advice, both face to face and in writing, or they may warn a duty holder that, in their opinion, the duty holder is failing to comply with the law. (For more on the term 'duty holder' see page 3.)

In carrying out the HSE's enforcement role, inspectors appointed under the HSW Act can:

- enter premises at any reasonable time, accompanied by a police officer if necessary
- examine, investigate and require the premises to be left undisturbed

- take samples, photographs and, if necessary, dismantle and remove equipment or substances

- review relevant documents or information such as risk assessments, accident books, or similar

- seize, destroy or make harmless any substance or article

- issue enforcement notices and start prosecutions.

An inspector may serve one of three types of notice:

- a prohibition notice tells the duty holder to stop an activity immediately

- an improvement notice sets out action needed to remedy a situation and gives the duty holder a date by which they must complete the action

- a Crown notice is issued under the same circumstances that would justify a prohibition or improvement notice, but is only served on duty holders in Crown organisations such as government departments, the Forestry Commission or the Prison Service.

SmartScreen Unit 301
Worksheet 2

THE REQUIREMENT FOR CONTROL OF RISKS

The control of risks is essential to securing and maintaining a healthy and safe workplace which complies with the relevant legal requirements. Most of the key elements required for effective health and safety management are similar to those required for other areas of successful organisations, such as finance or production.

The HSW Act requires employers to provide health and safety systems and procedures in the workplace, and directors and managers of any company that employs more than five employees can be held personally responsible for failure to control health and safety. The aim of the Management of Health and Safety at Work Regulations 1999 is to give more detail on how to achieve this and to encourage a more systematic and better organised approach to dealing with health and safety in all workplaces.

The control of risks is essential to the provision and maintenance of a safe and healthy workplace and ensures compliance with Regulation 3 of the Management of Health and Safety at Work Regulations. Risk assessment starts with the need for **hazard** identification.

The process of **risk** assessment is considered in more detail in Learning outcome 3.

The work operations that may be undertaken in the electrotechnical field will all have particular risks, and these and the relevant health and safety legislation are considered in more detail below.

Assessment criteria

1.2 Specify particular health and safety risks which may be present and the requirements of current health and safety legislation for the range of electrotechnical work operations

Hazard

A hazard is usually defined as anything with the potential to cause harm (eg chemicals, a fault on electrical equipment, working at height).

Risk

Risk is the chance (large or small) of harm actually being done when things go wrong (eg risk of electric shock from faulty equipment).

PREPARATION AND PLANNING

Preparation can be classed as the gathering of the relevant and necessary resources (materials, labour). Planning can be considered as the method of 'how and when' the resources will be used to perform the activity or complete the task.

Refurbishment work in buildings presents the greatest risk and must be planned, managed and monitored to ensure that workers are not exposed to risk from electricity during electrical and non-electrical work. Those responsible for planning and managing electrical refurbishment or installation work must have a complete understanding of the electrical system of the building in which the work takes place and liaise with the building owner/occupier.

INSTALLATION

Each year about 20 people die from electric shock or electric burns at work. Electrical installation work is often regarded as a single work process and workers are usually aware of the dangers of electricity, so this is clearly an area that must be given due consideration. However, because of differing work locations and a range of differing activities that are involved with electrical installation, there are risks associated with those work activities.

The most common categories of risk and causes of accidents at work when carrying out electrical installation are:

- electricity
- slips, trips and falls
- manual handling
- use of hand tools, including portable electric hand tools.

These will now be dealt with in more detail, but it should be recognised that other causes of accident may occur in the workplace such as:

- fire
- mechanical handling
- storage of goods and materials.

ASSESSMENT GUIDANCE

Battery-operated tools are much safer as the operating voltage, which is generally a maximum of 18 V or 24 V d.c., cannot normally, if faulty, give a fatal shock.

Electricity

BS 7671:2008 Requirements for Electrical Installations, which are non-statutory regulations, relate principally to the design and erection of electrical installations so as to provide for safety and proper functioning, for the intended use. If an installation is constructed in accordance with BS 7671, protection should be afforded to the user from:

- electric shock
- excessive temperature
- under-voltage, overvoltage and electromagnetic disturbances
- power supply interruptions
- arcing and burning.

However, a great number of electrical accidents occur because people are working on or near equipment that is thought to be dead but which is in fact live, or the equipment is known to be live and those involved in the work do not have adequate training or appropriate equipment, or they have not taken adequate precautions.

EAW Regulation 4(3) requires that every work activity, including operation, use and maintenance of a system and work near a system, shall be carried out in such a manner as not to give rise, so far as is reasonably practicable, to danger.

Housekeeping

This is one of the most important single items influencing safety within the workplace. Poor housekeeping not only causes an increase in the risk of fire, slips, trips and falls, but may also expose members of the public to risks created during building services engineers' work, such as fault diagnosis and maintenance, as these activities are normally carried out in occupied buildings. The following are some examples of good housekeeping.

- Stairways, passages and gangways should be kept free from materials, electrical power leads and obstructions of every kind.
- Materials and equipment should be stored tidily so as not to cause obstruction and should be kept away from the edges of buildings, roofs, ladder access, stairways, floor openings and rising shafts.

A badly managed site will inevitably lead to accidents

Tools should be kept neatly in a tool belt or bag

- Tools should not be left where they may cause tripping or other hazards. Tools not in use should be placed in a tool belt or tool bag and should be collected and stored in an appropriate container at the end of each working day.
- Working areas should be kept clean and tidy. Scrap and rubbish must be removed regularly and placed in proper containers or disposal areas.
- Rooms and site accommodation should be kept clean. Soiled clothes, scraps of food, etc should not be allowed to accumulate, especially around hot pipes or electric heaters.
- The spillage of oil or other substances must be contained and cleaned up immediately.
- All flammable liquids, liquified petroleum gas (LPG) and gas cylinders must be stored properly.

Gas bottles should be stored properly

Slips, trips and falls

These are the most common hazards to people as they walk around the workplace. They make up over a third of all major workplace injuries. Over 10 000 workers suffered serious injury because of a slip or trip in 2011. Listed below are the main factors that can play a part in contributing to a slip- or trip-free environment.

Flooring

- The workplace floor must be suitable for the type of work activity that will be taking place on it.
- Floors must be cleaned correctly to ensure they do not become slippery, or be of a non-slip type that it keeps its slip-resistance properties.
- Flooring must be fitted correctly to ensure that there are no trip hazards and any non-slip coatings must be correctly applied.
- Floors must be maintained in good order to ensure that there are no trip hazards such as holes, uneven surfaces, curled-up carpet edges, or raised telephone or electrical sockets.

- Ramps, raised platforms and other changes of level should be avoided. If they cannot be avoided, they must be highlighted.

Stairs

Stairs should have:

- high-visibility, non-slip, square nosings on the step edges
- a suitable handrail
- steps of equal height
- steps of equal width.

Contamination

Most floors only become slippery once they become contaminated. If **contamination** can be prevented, the slip risk can be reduced or eliminated.

Contamination of a floor can be classed as anything that ends up on a floor, including rainwater, oil, grease, cardboard, product wrapping, dust, etc. It can be a by-product of a poorly controlled work process or be due to bad weather conditions.

Contamination

The introduction of a harmful substance to an area.

Cleaning

Cleaning is important in every workplace; nowhere is exempt. It is not just a subject for the cleaning team. Everyone's aim should be to keep their workspace clear and deal with contamination such as spillages as soon as they occur.

The process of cleaning can itself create slip and trip hazards, especially for those entering the area being cleaned, including those undertaking the cleaning. Smooth floors left damp by a mop are likely to be extremely slippery. Trailing wires from a vacuum cleaner or polishing machine can present an additional trip hazard.

As people often slip on floors that have been left wet after cleaning, access to smooth wet floors should be restricted by using barriers, locking doors or cleaning in sections. Signs and warning cones only warn of a hazard, they do not prevent people from entering the area. If the water on the floor is not visible, signs and cones are usually ignored.

Human factors

How people act and behave in their work environments can affect the risk of slips and trips. For example:

- having a positive attitude toward health and safety. Dealing with a spillage instead of waiting for someone else to deal with it, can reduce the risk of slip and trip accidents
- wearing the correct footwear can also make a difference

ACTIVITY

It is very easy to trip or slip, which can result in an injury. What action should you take if you find that a carpenter has left an unprotected missing floorboard in a passageway?

- lack of concentration and distractions, such as being in a hurry, carrying large objects or using a mobile phone, all increase the risk of an accident.

Environmental factors

Environmental issues can affect slips and trips. The following points give an indication of these issues.

- Too much light on a shiny floor can cause glare and stop people from seeing hazards, for example, on floors and stairs.
- Too little light will prevent people from seeing hazards on floors and stairs.
- Unfamiliar and loud noises may be distracting.
- Rainwater on smooth surfaces inside or outside a building may create a slip hazard.

Footwear

Footwear can play an important part in preventing slips and trips. The following highlights some relevant points.

- Footwear can perform differently in different situations, for example, footwear that performs well in wet conditions might not be suitable where there are food spillages.
- A good tread pattern on footwear is essential on fluid-contaminated surfaces.
- Sole tread patterns should not be allowed to become clogged with any waste or debris on the floor, as this makes them unsuitable for their purpose.
- Sole material type and hardness are key factors influencing safety.
- The choice of footwear should take into account factors such as comfort and durability, as well as obvious safety features such as protective toecaps and steel mid-soles.

Manual handling techniques

Manual handling is one of the most common causes of injury at work and causes over a third of all workplace injuries. Manual handling injuries can occur almost anywhere in the workplace and heavy manual labour, awkward postures and previous or existing injury can increase the risk. Work-related manual handling injuries can have serious implications for both the employer and the person who has been injured.

The introduction of the Manual Handling Operations Regulations 1992 saw a change from reliance on safe lifting techniques to an analysis, using risk assessment, of the need for manual handling. The regulations established a clear hierarchy of manual handling measures:

Manual handling

The movement of items by lifting, lowering, carrying, pushing or pulling by human effort alone.

- avoid manual handling operations so as far as is reasonably practicable by re-engineering the task to avoid moving the load or by mechanising the operation

- if manual handling cannot be avoided, a risk assessment should be made

- reduce the risk of injury so as far as is reasonably practicable either by the use of mechanical aids or by making improvements to the task (for example by using two persons), the load and the environment.

How to manually handle loads using mechanical lifting aids

Even if mechanical handling methods are used to handle and transport equipment or materials, hazards may still be present in the four elements that make up the mechanical handling: the handling equipment, the load, the workplace and the human element.

- The handling equipment – mechanical handling equipment must be suitable for the task, well maintained and inspected on a regular basis.

- The load – the load needs to be prepared in such a way as to minimise accidents, taking into account such things as security of the load, flammable materials and stability of the load

- The workplace – if possible, the workplace should be designed to keep the workforce and the load apart

- The human element – employees who use the equipment must be properly trained.

Use of hand tools including hand-held power tools

The responsibilities of users of work equipment are covered in Section 7 of the HSW Act and Regulation 14 of the Management Regulations. Work equipment includes hand tools and hand-held power tools.

Section 7 of the HSW Act requires employees to take reasonable care of themselves and of others who may be affected by their acts or omissions at work and to cooperate with the employer to enable the employer to discharge his duties under the Act.

Regulation 14 of the Management Regulations requires employees to use equipment properly in accordance with instructions and training and to report to the employer any dangerous situations that might arise during the course of their work.

> **ASSESSMENT GUIDANCE**
>
> When you are manually handling a load, always make sure the route you will be taking is clear and all obstacles are removed. Never try to lift more than you can handle. Use safe handling techniques.

The hand tools that are most commonly used in electrical installation work are: screwdrivers, pliers, side cutters, hacksaw, adjustable spanners, hammers, chisels, cold chisels, files, centre punches, wire strippers, cable strippers, crimpers, ratchet hand, scissors, etc.

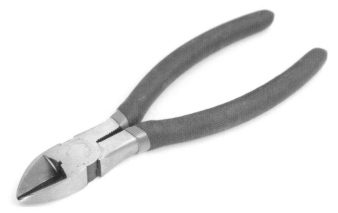

Side cutters

Bolster chisel and lump hammer

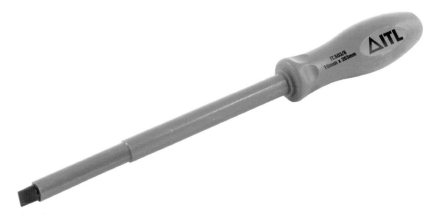

Insulated screwdriver (courtesy of Insulated Tools Ltd)

The most commons injuries from the use of hand tools are:

- blows and cuts to the hands or other parts of the body
- eye injuries due to the projection of fragments or particles
- sprains due to very abrupt movements or strains
- electric shock and/or burns from contact with live conductors.

ACTIVITY

How should a mushroomed head be removed from a cold chisel?

The principal causes of injury are:

- inappropriate use of the tools
- use of faulty or inappropriate tools
- use of poor quality tools
- not using personal protection equipment
- forced postures.

Preventive measures:

- use good quality tools in accordance with the type of work being carried out
- ensure employees are trained in the correct use of tools
- use approved insulated hand tools (IEC 60900) when working in the vicinity of live parts
- use protective goggles or glasses (BS EN 166) in all cases and above all when there is a risk of projected particles
- use gloves to handle sharp tools
- periodically check tools (repair, sharpening, cleaning, etc)
- periodically check the state of handles, insulating coverings, etc
- store and/or transport tools in boxes, tool bags or on suitable panels, where each tool has its place.

Hand-held power tools (portable electrical equipment)

Almost 25% of all reportable accidents involve portable electrical equipment, the majority being caused by electric shock. Electrical equipment that is hand held or being handled when switched on presents a high level of risk to the user if it does develop a fault.

Damaged transformer and poorly maintained equipment lead to a high risk of injury

ASSESSMENT GUIDANCE

Although battery-powered tools have no trailing leads, the chargers do and have a 230 V or 110 V connection. It may be better to have a central charging area where all leads can be kept under control.

ACTIVITY

Name two advantages of using battery-powered tools compared to mains supplied tools.

The principal causes of accidents are:

- use of damaged, defective or unsuitable equipment
- lack of effective maintenance
- misuse of equipment
- 'unauthorised' equipment (eg electric heaters, kettles, coffee percolators, electric fans) being used by employees
- use of power tools in a harsh environment (construction sites, etc).

Preventive measures:

- correct selection and use of equipment
- ensuring employees are trained in the correct use of tools
- effective maintenance regime for all power tools, including user checks
- keeping of test records following portable appliance testing
- removal of damaged equipment from use.

The particular legal requirements relating to the use and maintenance of electrical equipment are contained in the following:

- **Section 2 of the HSW Act** requires 'the provision and maintenance of plant … (that is) … so far as is reasonably practicable, safe'.
- **Electrical Equipment (Safety) Regulations 1994** require certain safety objectives to be met, including design and construction, to ensure protection against hazards arising from the electrical equipment.
- **Supply of Machinery (Safety) (Amendment) Regulations 1994** contain a general requirement for protection against electrical hazards.
- **Management of Health and Safety at Work Regulations 1999** require employers to carry out risk assessment, make arrangements to implement necessary measures, appoint competent people and arrange for appropriate information and training.
- **Provision and Use of Work Equipment Regulations 1998 (PUWER)** require the employer (person in control) to select suitable work equipment (Regulation 5) and to 'ensure that work equipment is maintained in an efficient state'.
- **Electricity at Work Regulations 1989 (EAW)** require people in control of electrical systems to ensure they are safe to use and maintained in a safe condition.
- **EAW Regulation 4(2)** requires that all systems be maintained, so far as reasonably practicable, to prevent danger. This requirement covers all items of electrical equipment including fixed, portable and transportable equipment.

HOW TO CONDUCT A VISUAL INSPECTION OF PORTABLE ELECTRICAL EQUIPMENT

To maintain the safety and integrity of tools and equipment, regular in-service inspection and testing should be undertaken to confirm that the equipment remains in a safe condition. Portable appliance testing (PAT) is the term used to describe the examination of electrical appliances and equipment to ensure they are safe to use. There are three categories of in-service inspection or testing for portable tools and equipment:

- user checks (pre-use inspections) – users play an important role by checking equipment before use for signs of damage or obvious defects liable to affect safety in use
- periodic formal inspection or checks
- periodic combined inspection and testing.

The user checks are a visual check only, and should deal with the following inspection requirements separately for the equipment, supply lead and plug.

Equipment

- Equipment should be manufactured to relevant standards (BS or BS EN).
- The casing should have no visible damage and be free from dents and cracks.
- The switches should operate correctly.
- The supply lead should be secure and correctly connected.
- It should be suitable for the task.

Supply leads

- Supply leads should be manufactured to relevant standards (BS or BS EN).
- They should be suitable for the environment.
- They should be free from cuts or fraying.
- There should be no visible exposed conductor insulation (damaged sheath) or exposed live conductors.
- There should be no signs of damage to the cord sheath.
- There should be no joints evident.

Supply lead plugs

- Supply lead plugs should be manufactured to relevant standards (BS or BS EN).
- The insulation should have no damage.
- The cord/cable connections should be correct and secure.

ACTIVITY

The flex used to supply portable equipment on a construction site is liable to mechanical damage. Use wholesalers' catalogues or websites to select suitable flexible cables.

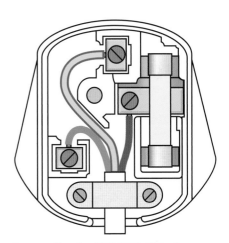

A correctly wired BS 1363 13 A plug

ASSESSMENT GUIDANCE

Ordinary 13 A plugs are not suitable for use outdoors or in damp situations. Special types should be used.

ACTIVITY

What are the colours used to identify plugs and sockets for the following voltages?

- 110 V single-phase
- 230 V single-phase
- 400 V three-phase

The user checks must be backed up by periodic inspection and, where appropriate, testing. At that time, a thorough examination is undertaken by a nominated person, competent for that purpose. Schedules giving details of inspection and maintenance periods, together with records of the inspection, should form part of the procedure.

What to do when portable electrical equipment fails visual inspection

Most electrical safety defects can be found by visual examination, but some types of defect can only be found by testing. However, it is essential to understand that visual examination is an essential part of the process because some types of electrical safety defect cannot be detected by testing alone.

A relatively brief user check, based upon simple training and the use of a brief checklist, is a very useful part of any electrical maintenance regime. If the user checks detailed on page 19 are carried out, 95% of all faults will be identified and the appropriate action taken. No record is needed if there are no defects found. However, if equipment is found to be unsafe, it must be removed from service immediately. It must be labelled to show that it must not be used and the fault must be reported to a responsible person.

TERMINATIONS AND CONNECTION

The connection between a conductor and other equipment (distribution boards, fixed appliances, etc) or accessory is normally called a 'termination'. Connections between conductors are usually termed 'joints'.

A poorly constructed joint or termination may give rise to arcing with the passage of load current, resulting in localised heating (and an increased fire risk) and progressive deterioration of the joint or termination, possibly leading to eventual failure and an open circuit.

When a fault occurs elsewhere on an associated part of an electrical **system**, a poor joint or termination may constitute a hazard by limiting the flow of fault current, thereby causing delayed operating of the circuit protective device.

In an extreme case, the passage of high-fault current may cause the defective joint to burn into an open-circuit fault before the circuit protective device has time to operate, with the result that the original fault remains uncleared. This is particularly hazardous in the case of an earth fault, since exposed conductive parts may become and remain live at the system voltage, giving an increased risk of electric shock.

This risk is identified in both the Electricity at Work Regulations and the IET Wiring Regulations.

System

This covers all and any electrical equipment which is, or may be, connected to an electrical energy source, and includes that source.

ACTIVITY

Can you identify three requirements or good practices when terminating conductors into a terminal?

EAW Regulation 10 states: 'Where necessary to prevent danger, every joint and connection in a system shall be mechanically and electrically suitable for use.' This regulation applies to every joint and connection, whether temporary or permanent. It therefore covers not just in-line cable joints but also plugs and sockets, connectors, terminal blocks and the termination/connection of cables to equipment. The regulation applies to all conductors, ie protective as well as circuit conductors, as all joints and connections must be capable of carrying fault current as well as normal load current. This is an absolute requirement. The phrase 'so far as is reasonably practicable', does not appear in the regulation.

This requirement is restated in Section 526 of the IET Wiring Regulations: 'Every connection between conductors or between a conductor and other equipment shall provide durable electrical continuity and adequate mechanical strength and protection.'

Section 526 also requires the following:

- The means of connection shall take account of the number and shape of the wires forming the conductor, the cross-sectional area of the conductor and the number of conductors to be connected together.
- The temperature at the terminals shall not impair the effectiveness of the conductor insulation.
- Every connection shall be accessible for inspection, testing and maintenance.
- There shall be no mechanical strain on the conductor connections.

INSPECTION, TESTING AND COMMISSIONING

The importance of regular inspection and testing of electrical installations cannot be overstated, but unfortunately it is an aspect of electrical safety which is very often ignored by those in control of premises. All electrical installations deteriorate due to a number of factors, ie damage, wear and tear, corrosion, excessive electrical loading, ageing and environmental influences. Electrical installations that are inspected after the completion of any work will achieve the following aims:

- the installation is safe to use
- the installation complies with the requirements of the IET Wiring Regulations
- the installation complies with the project specification (commissioning).

A badly installed electrical system presents a number of risks both to the users of the system and to persons who may be required to undertake modification or alteration to the system in the future.

ASSESSMENT GUIDANCE

Temporary installations should be installed to the same high standards. Just because they are used for a few months or so does not mean standards can be dropped.

ACTIVITY

Which one of the following is the stage of an electrical installation where inspection and testing is carried out?

a) during the installation
b) completion of the installation
c) both during and at completion
d) following the warranty period

EAW Regulation 4(1) states: 'All systems shall at all times be of such construction as to prevent, "so far as is reasonably practicable", danger.' The overall requirement in Regulation 4(1) aims to ensure that the electrical system can fulfil the user's requirements safely:

- It must be adequate for all anticipated normal load conditions.
- It must ensure that any fault currents (whether short circuit or earth fault currents) are disconnected in such a way as to avoid damage to the system and injury to people.

It should be noted that Regulation 4(1) covers the design of a **system** as well as its physical construction and also the correct selection of equipment that makes up the **system**.

In order to make sure that inspection testing and commissioning work is carried out satisfactorily, the inspection and test procedure must be carefully planned and carried out and the results correctly documented.

The certificate and test results will normally be issued to those persons who ordered the tests to be undertaken and usually in the format shown in Appendix 6 of the IET Wiring Regulations – Model Forms for Certification and Reporting.

Part 6 of the IET Wiring Regulations states that every electrical installation shall, either during construction, on completion, or both, be inspected and tested to verify, 'so far as is reasonably practicable', that the requirements of the Regulations have been met.

In carrying out such inspection and test procedures, precautions must be taken and the following safety factors should be considered:

- safe working practices to be agreed
- safety precautions must be in place before testing commences
- all persons who may be affected by the testing should be notified
- all information relevant to the installation should be obtained (previous test results, circuit diagrams, details of sensitive equipment)
- relevant circuits and equipment are identified
- the correct isolation procedures are adopted (EAW R14)
- only competent persons should be engaged in inspection and testing (EAW R16)
- test instruments should be suitable and calibrated.

More detail on inspection testing, commissioning and record keeping is contained in 2357 Unit 307.

FAULT DIAGNOSIS AND RECTIFICATION

Diagnosing and fault finding in an electrical installation is a difficult task that needs knowledge, experience and a systematic approach. Fault situations are rarely, if ever, the same, and each problem has to be investigated on an individual basis. More complex installations mean more complex faults and more complex faults means more complex solutions! However, the adoption of a regular systematic inspection will ensure procedure faults can be identified and rectified before injury to persons or damage to equipment occurs.

Fault diagnosis and rectification is covered in more detail in 2357 Unit 308.

Maintenance

Maintenance can be broadly categorised as unplanned and planned.

- **Unplanned maintenance** is where tasks are undertaken as a reaction to a situation.
- **Planned maintenance** is organised, controlled and reviewed.

Planned/scheduled maintenance

Preventive maintenance is carried out before a failure or malfunction occurs, normally at predetermined intervals. Preventive maintenance is the normal practice with electrical systems and equipment. Its aim is to prevent danger arising 'so far as is reasonably practicable'. Preventive maintenance is coupled with maintenance to deal with any unexpected failure that cannot be left unattended. There are two types of planned maintenance:

- **Corrective maintenance** is work done to restore plant or equipment to correct operation, ie within acceptable parameters.
- **Predictive maintenance** is maintenance carried out at predetermined intervals of time, or after a certain number of operations, etc.

Unplanned maintenance

Unplanned maintenance is reactive and normally only takes place after a failure. There are two types of unplanned maintenance.

- **Breakdown maintenance** is carried out where equipment failure or malfunction does not result in danger and the consequences of the failure are otherwise acceptable.
- **Emergency maintenance**, applied to electrical systems, describes the work that must be put in hand immediately in order to rectify an unexpected dangerous condition, such as might result from damage to equipment.

EAW Regulation 4(2) recognises that the integrity of any electrical system can only be preserved in its 'as initially installed' condition if it is subsequently maintained and repaired as necessary, on a regular on-going basis. This is no more or less than what one would expect with the use of any item of electrical or mechanical plant.

This regulation embraces all electrical systems, installations and equipment in the workplace, whether fixed, movable, hand-held, or transportable. The obligation to maintain arises only if danger would otherwise result. Danger here means risk of death or personal injury from electric shock, electric burn, electrical explosion or arcing, or from fire or explosion initiated by electrical energy.

The EAW Regulations do not lay down frequencies for maintenance; these are for the **duty holder** to decide. HSE's guidance on Regulation 4(2) states that 'the quality and frequency of maintenance should be sufficient to prevent danger "so far as is reasonably practicable"'.

The Provision and Use of Work Equipment Regulations 1998 (PUWER) apply to any machinery, appliance, apparatus or tool used for carrying out any work activity, whether or not electricity is involved. Regulation 5 requires such machinery, etc to be maintained in an efficient state, in efficient working order and in good repair.

Additionally, any maintenance logs for machinery shall be kept up to date and Regulation 22 places a general duty on employers to take appropriate measures to ensure that maintenance operations can be carried out safely. Such measures would include the siting of electrical equipment, taking account of maintenance needs.

Regulation 5 of the Workplace Regulations requires employers to maintain workplaces and associated equipment, devices and systems in an efficient state, in efficient working order and in good repair.

ASSESSMENT GUIDANCE

Regular maintenance is important to keep machinery in optimum condition and to prevent undue wear and tear. Remember the old saying 'A stitch in time saves nine'. Topping up the oil could save thousands of pounds of damage later.

Know the procedures for dealing with health and safety in the work environment

THE ROLE OF SAFETY CULTURE

The Health and Safety Commission's (HSC) commonly accepted definition of safety culture is: 'The product of individual and group values, attitudes, perceptions, competencies, and patterns of behaviour that determine the commitment to, and the style and proficiency of, an organisation's health and safety management.'

An organisation's culture can have as big an influence on safety outcomes as the safety management system itself; with safety culture being a subset of the overall organisational or company culture. Poor safety culture has contributed to many major incidents (Piper Alpha oil platform in the North Sea; the fire at Kings Cross underground station; the sinking of the Herald of Free Enterprise passenger ferry; the passenger train crash at Clapham Junction; Chernobyl) and personal injuries. Success in this area normally comes from good leadership, good worker involvement and good communications.

By paying attention to human factors, forward-looking companies can identify and deal with potential hazards before they manifest themselves as accidents. This coupled with legislation in the form of regulations and Approved Codes of Practice (ACoP) that are easily understood, and hence complied with, can have a positive effect on health and safety standards and help prevent or reduce accidents and incidents.

PROCEDURES FOR HANDLING INJURIES SUSTAINED ON SITE

The type of accident that can occur in the workplace is dependent on the work activity being undertaken but can range from a cut finger to a **fatality**, or from a vehicle collision to the collapse of a structure. The person in control of the premises needs to be prepared to deal with all types of accidents to ensure that the injured person can be treated quickly and effectively and that all the legal obligations are met.

Having a well-established procedure that everyone on site is aware of and understands will enable the person in control of premises to cope calmly and effectively when dealing with an accident. Good management following an accident will ensure that the injured person is attended to promptly, appropriate records are made, the accident is reported correctly and any lessons to be learned from the accident are understood and communicated to the workforce.

Assessment criteria

2.1 State the procedures that should be followed in the case of accidents which involve injury, including requirements for the treatment of electric shock/electrical burns.

Fatality

Death.

ACTIVITY

How should you deal with a deeply cut hand sustained while cutting a piece of metal trunking?

The procedures to be followed in the event of any accident or incident should be clear and specific to the project or site and should detail the following as a minimum:

- name of the appointed person(s) who will take control when someone is injured or falls ill
- name of the person(s) who will administer first aid
- location of the first-aid boxes and name of the person(s) responsible for ensuring they are fully stocked
- course of action that must be followed by the appointed person who takes control in the event of an accident
- guidance on action to take after the accident
- how information should be recorded and by whom (F2508 RIDDOR Form, which can be found on the HSE website www.hse.gov.uk/forms/incident/).

HOW TO DEAL WITH ELECTRIC SHOCKS

If all of the correct requirements are met, precautions taken and training of staff undertaken, it is unlikely that an electrical accident will occur. However, procedures should be in place to deal with electric shock injury in the event of an accident. The recommended procedure for dealing with a person who has received a low voltage shock is as follows.

- Raise the alarm (colleagues and a trained first-aider).
- Make sure the area is safe by switching off the electricity supply.
- Request colleagues to call an ambulance (999 or 112).
- If it is not possible to switch off the power supply, move the person away from the source of electricity by using a non-conductive item.
- Check if the person is responsive, whether their airway is clear and whether they are breathing.
- If the person is unconscious but breathing, move them into the recovery position.
- If they are unconscious but not breathing, start to give cardiopulmonary resuscitation (CPR):
  - CPR is undertaken by interlocking the hands and giving 30 chest compressions in the centre of the chest, between the two pectoral muscles, at a rate of about 100 pulses per minute.
  - Tilt the casualty's head back gently, by placing one hand on the forehead and the other under the chin, to open the airway and give two mouth-to-mouth breaths.
  - Repeat the cycle of 30 compressions to two breaths until either help arrives or the patient recovers.
- Any minor burns should be treated by placing a sterile dressing over the burn and securing with a bandage.

PROCEDURES FOR RECORDING ACCIDENTS AND NEAR MISSES AT WORK

An accident is defined by the HSE as, 'any unplanned event that results in injury or ill health of people, or damage or loss to property, plant, materials or the environment, or a loss of a **business opportunity**'.

A **near miss** is an unplanned event that does not result in injury, illness or damage, but had the potential to do so. Normally only a fortunate break in the chain of events prevents an injury, fatality or damage taking place. So a near miss could be defined as any incident that could have resulted in an accident. The keeping of information on near misses is very important in helping to prevent accidents occurring. Research has shown that damage and near miss accidents occur much more frequently than injury accidents and therefore give an indication of hazards.

The Social Security Act 1975 specifically requires employers to keep information on accidents. This should be the Statutory Accident Book for all Employee Accidents (Form B1510, found on the HSE website www.hse.gov.uk/forms/incident/) or equivalent. Each entry should be made on a separate page and the completed page securely stored to protect personal data (under the Data Protection Act 2003). An entry may be made by the employee or by anyone acting on their behalf. This information should be kept for a period of not less than three years.

Business opportunity

In this context, the opportunity to make profit from the work or contract.

Near miss

Any incident that could, but does not, result in an accident.

Employers are required to keep a record of all accidents in the Statutory Accident Book, or an equivalent

Reporting the incident

The reporting of certain types of injury and incidents to the enforcing authority (the HSE or the local authority) is a legal requirement under the Reporting of Injuries, Diseases and Dangerous Occurrences Regulations 2013 (RIDDOR). Failure to comply with these regulations is a criminal offence.

RIDDOR states that deaths, specified injuries (listed in the regulations) and injuries resulting in absence from work for over seven days, and dangerous occurrences (again listed in the regulations) and occupational diseases must be reported. This seven-day period does not include the day of the accident, but does include weekends and rest days. The report must be made within 15 days of the accident. It is the responsibility of employers or the person in control of premises to report these types of incidents.

Reportable specified injuries include fractures, amputations, permanent loss of sight, crush injuries and serious burns. A dangerous occurrence is a 'near-miss' event (incident with the potential to cause harm). There are also special requirements for gas incidents. Accidents must be recorded, but not reported, where they result in a worker being incapacitated for more than three consecutive days.

ACTIVITY

An operative complains that they have something in their eye. What action should be taken?

The police and HSE have the right to investigate fatal accidents at work. Therefore all fatal accidents must also be notified to the police. The police will often notify the HSE, but it is always a sensible precaution to ensure that the HSE has been notified.

Investigating accidents

There is nearly always something to be learned following an accident and ideally the causes of all accidents should be established regardless of whether injury or damage resulted. The level and nature of an investigation should reflect the significance of the event being investigated. The results of the accident investigation may lead to a review, possible amendment to the risk assessment and appropriate action to prevent similar accidents from occurring.

Keeping records

There are numerous records to keep following even a minor accident. Easily accessible records should be maintained for all accidents that have occurred. In addition to the legal requirements, accident information can help an organisation identify key risk areas within the business. The accident book must be kept for three years following the last entry.

- The F2508 for reportable incidents should be kept for a period of not less than three years from the date the accident occurred.
- Appointed trade union representatives and employee safety representatives should be provided with a copy of the F2508 if it relates to the workplace or the group of employees represented.
- It is advisable to keep copies of any accident investigation reports for the same period as above (three years).

WHAT TO DO IN AN ACCIDENT OR EMERGENCY

The HSE definition of an accident has been given in the section 'Procedures for recording accidents and near misses at work' on page 27.

Emergency procedures

Where an emergency situation occurs and the emergency services need to be called, be sure you know the following information:

- the address and location of the incident
- the nature of the incident
- any difficulties the emergency services may encounter whilst trying to get to the location within the site
- any immediate dangers such as explosive materials, persons trapped etc.

Should an emergency occur that requires evacuation, it is essential you know what or where the designated escape route is. On construction sites, due to their ever-changing nature, designated escape routes may frequently change so always be sure you are familiar with them. What may have been a safe escape route one day could be a dangerous route the next.

In nearly all fire situations, the emergency services should be called, but some very small fires may be extinguished before they become too serious. Be sure to know how to tackle different fires and know where appropriate fire-fighting devices are located.

Emergencies

Emergency procedures are there to limit the damage to people and property caused by an incident. Although the most likely emergency to be dealt with is fire (see pages 57 and 68), there are many more emergency situations that need to be considered, including the following.

Electrical fire or explosion

Fires involving electricity are often caused by lack of care in the maintenance and use of electrical equipment and installations. The use of electrical equipment should be avoided in potentially flammable atmospheres as far as is possible. However, if the use of electrical equipment in these areas cannot be avoided, then equipment purchased in accordance with the Equipment and Protective Systems Intended for Use in Potentially Explosive Atmospheres Regulations 1996 must be used.

> **ASSESSMENT GUIDANCE**
>
> Whether you are being assessed on site or not, always wear your PPE.

Escape of toxic fumes or gases

Some gases are poisonous and can be dangerous to life at very low concentrations. Some toxic gases have strong smells such as the distinctive 'rotten eggs' smell of hydrogen sulphide (H_2S). The measurements most often used for the concentration of toxic gases are parts per million (ppm) and parts per billion (ppb). More people die from toxic gas exposure than from explosions caused by the ignition of flammable gas. With toxic substances, the main concern is the effect on workers of exposure to even very low concentrations. These could be inhaled, ingested (swallowed) or absorbed through the skin. Since adverse effects can often result from cumulative, long-term exposure, it is important not only to measure the concentration of gas, but also the total time of exposure.

> **KEY POINT**
>
> Toxic substances can stop one or more organs in the human body working. Mercury, lead and carbon monoxide, for example, are all toxic.

Gas explosion

A gas explosion is an explosion resulting from a gas leak in the presence of an ignition source. The main explosive gases are natural gas, methane, propane and butane because they are widely used for heating purposes in temporary and permanent situations. However,

Combustible

Able to catch fire and burn easily.

many other gases, such as hydrogen, are **combustible** and have caused explosions in the past. The source of ignition can be anything from a naked flame to the electrical energy in a piece of equipment.

Industrial gas explosions can be prevented with the use of intrinsic safety barriers to prevent ignition. The principle behind intrinsic safety is to ensure that the electrical and thermal energy from any electrical equipment in a hazardous area is kept low enough to prevent the ignition of flammable gas. Items such as electric motors would not be permitted in a hazardous area.

Assessment criteria

2.3 State the limitations of their responsibilities in terms of health and safety in the workplace.

EMPLOYER AND EMPLOYEE RESPONSIBILITIES

Employers have a duty of care to each of their employees. This duty rests solely with the employer and cannot be assigned to others, such as consultants who may be retained to offer advice on health and safety matters, or sub-contractors who are employed to undertake tasks within the company.

All organisations should have a clear policy for the management of health and safety. The policy sets the direction for health and safety within the organisation, and its contents need to be clearly communicated to everyone within the organisation to ensure everyone understands what their responsibilities are in day-to-day operations. Everyone has responsibility for safety, but under the law there is no equality of responsibility between those who provide direction and create policy (the principle or employer) and those who are employed to follow the policy (employees).

Employers have the following duties under the HSW Act:

- the health, safety and welfare at work of employees, and other workers whether they are contractors, casual, temporary, part time or trainees
- the health and safety of anyone who is allowed to use the organisation's equipment
- the health and safety of anyone who may be affected by the organisation's activities, ie general public or adjacent organisations or neighbours.

Examples of matters under the control of an employer include:

- establishing policy to ensure that electrical equipment is purchased to an appropriate specification
- establishing policy to ensure that electrical equipment is properly maintained (including user inspection)
- implementing policy through the introduction of appropriate management systems

- on-going monitoring to confirm that the policy is properly implemented and remains fully relevant.

Employees have specific responsibilities under the HSW Act and these can be summarised as:

- to take reasonable care for the health and safety of themselves and others who may be affected by their acts or omissions at work

- to cooperate with their employer and others to enable them to fulfil their legal obligations

- not to interfere with or deliberately misuse anything provided in the interests of health and safety.

Examples of matters under the control of an employee include:

- adherence to company procedures and systems of work

- use of equipment in accordance with information and training provided

- not to use equipment that is faulty or damaged but to report it in accordance with company arrangements for dealing with defective equipment.

In summary, under the HSW Act employers have duties to ensure that appropriate management systems are established so that electrical work can be undertaken in a safe manner, whilst employees have a responsibility to comply fully with such management systems.

WHAT TO DO IF YOU HAVE CONCERNS ABOUT HEALTH AND SAFETY ISSUES

Employees are responsible for ensuring that the work they are required to undertake is carried out in a manner which is safe to themselves and other persons who may be affected by the work activity. They must undertake this work activity in accordance with the instruction or procedure provided by the employer.

If an employee has concerns about health and safety at work, or feels that there is a situation which they believe exceeds their level of responsibility, then these concerns should be raised with their supervisor or line manager. If the organisation has a recognised trades union and a safety representative has been appointed, it may be appropriate for the safety representative or the trades union official to be the first point of contact.

Assessment criteria

2.4 State the actions to be taken in situations which exceed their level of responsibility for health and safety in the workplace.

ASSESSMENT GUIDANCE

If you see someone carrying out a dangerous procedure, report it to your supervisor. It may not make you the most popular person but may save their, or another person's life, maybe even yours.

ACTIVITY

Are mines and quarries covered by BS 7671?

Assessment criteria

2.5 State the procedures that should be followed in accordance with the relevant health and safety regulations for reporting health, safety and/or welfare issues in the workplace.

WHAT TO DO IF YOU HAVE CONCERNS ABOUT YOUR WORK ACTIVITY

The law requires employers and the self-employed to conduct their business in such a way as to ensure, 'so far as is reasonably practicable', that persons affected are not exposed to risks to their health or safety. This includes providing essential welfare facilities for employees.

If it is believed that an employer's (or someone else's) work activity is creating a risk or causing damage to health, then the concern should be raised with that employer or person responsible for the work activity. If no improvement is made and the safety or health risk continues, then it should be reported to the relevant enforcing authority, which should be asked to investigate.

There are two relevant enforcing authorities.

The HSE is responsible for enforcing health and safety at workplaces that include:

■ mines
■ factories
■ building sites
■ nuclear installations
■ hospitals and nursing homes
■ gas, electricity and water systems
■ schools and colleges, offshore installations.

The local authority environmental health department is responsible for enforcing health and safety at workplaces that include:

■ shops
■ hotels
■ restaurants
■ leisure premises
■ nurseries and playgroups
■ public houses and clubs
■ offices (except government offices).

Assessment criteria

2.6 Specify appropriate responsible persons to whom health and safety and welfare related matters should be reported.

WHO TO REPORT CONCERNS TO

All companies need an effective internal system for reporting health-, safety- and welfare-related matters and this should be readily available to all employees and come complete with the names and contact details of the relevant responsible persons.

The names, positions and duties of those people within the organisation who have responsibility for health and safety will normally be contained within the safety policy. This will include:

- managers (directors, site managers, supervisors)
- specialists, health and safety advisors, safety officers, first aiders, fire officers)
- employee representatives (trades union representatives, safety representatives)
- Health and Safety Executive Officers and Local Authority Environmental Health Officers.

OUTCOME 3

Understand the procedures for establishing a safe working environment

Assessment criteria

3.1 State the procedure for producing risk assessments and method statements in accordance with their level of responsibility.

THE REQUIREMENT FOR RISK ASSESSMENTS

The HSW Act requires employers to provide health and safety systems and procedures in the workplace, and directors and managers of any company that employs more than five employees can be held personally responsible for failure to control health and safety. The general duties of Section 2 of the HSW Act imply that risk assessment is necessary. The aim of the Management of Health and Safety at Work Regulations 1999 (MHSWR) is to give more detail on how to achieve this and to encourage a more systematic and better organised approach to dealing with health and safety in all workplaces. The main objective of risk assessment is to determine the measures required by the organisation to comply with relevant health and safety legislation and so reduce the level of occupational injuries and ill health.

MHSWR Regulation 3 requires that: 'Every employer and self-employed person shall make a suitable and sufficient assessment of:

- the risks to the health and safety of his employees to which they are exposed whilst they are at work; and

- the risks to the health and safety of persons not in his employment arising out of or in connection with the conduct by him of his undertaking…'.

Risk assessment starts with the need for hazard identification.

A hazard is usually defined as anything with the potential to cause harm (chemicals, working at height, a fault on electrical equipment).

Risk is the chance (large or small) of harm actually being done when things go wrong (eg risk of electric shock from faulty equipment) which is usually evaluated in terms of:

- the likelihood of occurrence, ie is a hazard going to occur; and

- the severity of outcome, ie how serious the resulting injury will be.

All organisations must have the freedom to prepare systems of work that match the risk potential of their particular work activity and which are practical in their application. Electrotechnical work operations are no different to other work disciplines in this respect and employers must examine the work activities and the workplace and introduce

ASSESSMENT GUIDANCE

A risk assessment identifies the hazards and identifies methods of reducing the risk. It should be continuously reviewed as the working environment changes.

systems of work through a process of 'risk assessment' that takes account of the risks and the application of the health and safety legislation designed to manage those risks.

The HSE promotes a risk assessment process entitled **five steps to risk assessment**. The five steps are as follows:

- identify the hazards
- decide who may be harmed and how
- evaluate the risks and decide on precautions
- record the findings and implement them
- review the assessment and update if necessary.

The control of **risks** is essential to the provision and maintenance of a safe and healthy workplace and ensures **compliance** with the relevant legal requirements.

Risk assessment starts with the need for **hazard** identification. Risk assessment is usually evaluated in terms of:

- the likelihood of something happening (ie whether a hazard is going to occur)
- the severity of outcome (ie how serious the resulting injury will be).

To control these hazardous situations, duty holders need to:

- find out what the hazards are
- decide how to prevent harm to health
- provide control measures to reduce harm to health
- make sure the control measures are used
- keep all control measures in good working order
- provide information, instruction and training for employees and others
- provide monitoring and health surveillance in appropriate cases
- plan for emergencies.

CONTROL OF RISKS

If the risk level identified is high, the control measures may need to be more extensive. The aim of risk assessment is to reduce all residual risks to as low a level as reasonably practicable and use the resulting information to prepare control measures as necessary. Many hazards have specific regulations or guidance available to give advice on reducing the risk (lead, electricity, asbestos) and these should be referred to in the first instance, to obtain advice for the preparation of control measures.

Risk

The chance (large or small) of harm actually being done when things go wrong (eg risk of electric shock from faulty equipment).

Compliance

The act of carrying out a command or requirement.

Hazard

Anything with the potential to cause harm (eg chemicals, working at height, a fault on electrical equipment).

KEY POINT

You must be able to identify hazards as part of a risk assessment. It should then be possible to eliminate, reduce, isolate or control the risk by the application of suitable control measures. The use of PPE should be a last resort.

Assessment criteria

3.2 Describe the procedures for working in accordance with provided, pre-determined risk assessments, method statements and safe systems of work.

SAFE SYSTEMS AT WORK

A safe system of work is a work method that results from a systematic examination of the working process to identify the hazards and to specify work methods designed either to completely remove the hazards or control and minimise the relevant risks. Section 2 of the Health and Safety at Work etc Act 1974 (HSW Act) requires employers to provide safe plant and systems of work. Many of the regulations made under the HSW Act have more specific requirements for the provision of safe systems of work.

Safe systems of work should be developed by a **competent person**, that is, a person with sufficient training and experience or knowledge to assist with key aspects of safety management and compliance. Staff who are actively involved in the work process also have a valuable role to play in the development of the system. They can help to ensure that it is of practical benefit and that it will be applied diligently.

All safe systems of work need to be monitored regularly to ensure that they are fully observed and effective, and updated as necessary.

Safe systems of work are normally formal and documented, but may also be given as a verbal instruction. Examples of documented safe systems of work would be for asbestos removal, air-conditioning maintenance, working on live electrical equipment and portable appliance testing.

Method statements

A method statement is a description of a safe system of work, presenting in a logical sequence, and generally in writing, a method of undertaking an activity without risk to health. The method statement should be clear, and illustrated where applicable by simple sketches. Statements are for the benefit of those carrying out the work and their immediate supervisors and should not be overcomplicated. The statement includes all the risks identified in the risk assessment and the measures to control those risks. The statement need be no longer than necessary to achieve the objective of safe working.

Employees should know what to do if the work process has changed and the method statement no longer achieves its aim. Method statements must be reviewed on a regular basis to accommodate any such changes in the work process.

Permit-to-work procedures

A permit-to-work (PTW) procedure is a specialised written safe system of work that ensures that potentially dangerous work is done safely. Examples of such work include: work in confined spaces, **hot work**, work with asbestos-based materials, work on pipelines with hazardous contents, or work on high voltage electrical systems (above 1000 V) or complex lower voltage electrical systems.

Competent person

Recognised term for someone with the necessary skills, knowledge and experience to manage health and safety in the workplace.

ASSESSMENT GUIDANCE

Practise producing a method statement for a small electrical installation of your choice.

Hot work

Work that involves actual or potential sources of ignition and done in an area where there is a risk of fire or explosion (eg welding, flame cutting or grinding).

A PTW procedure also serves as a means of communication between site/installation management, plant supervisors and operators and those who carry out the hazardous work. Essential features of PTW systems are:

- clear identification of who may authorise particular jobs (and any limits to their authority) and who is responsible for specifying the necessary precautions
- a PTW should only be issued by a technically competent person, who is familiar with the system and equipment, and is authorised in writing by the employer to issue such documents
- training and instruction in the issue, use and closure of permits
- monitoring and **auditing** to ensure that the system works as intended
- clear identification of the types of work considered hazardous
- clear and standardised identification of tasks, risk assessments, permitted task duration and any additional activity or control measure that occurs at the same time.

The effective operation of a PTW system requires involvement and cooperation from a number of persons. The procedure for issuing a PTW should be written and adhered to.

THE HIERARCHY OF CONTROL

The control of risk is essential to the provision of a safe working environment that complies with all of the relevant legal requirements. To maintain this safe working environment, any control measures that are produced require assessment and if necessary amendment, or the introduction of new measures.

There is an acknowledged 'hierarchy of control', which is simply a list of measures designed to control risks, that are considered in order of importance, effectiveness or priority. This control sequence normally begins with an extreme measure of control and ends with personal protective equipment as a last resort:

1 *eliminate* the risk by designing out or changing the process
2 *reduce* the risk by substituting less hazardous substances
3 *isolate* the risk by using enclosures, barriers or worker segregation
4 *control* the risk by the introduction of guarding, local exhaust systems, reduced voltages or residual current devices
5 *management control*, such as safe systems of work and training
6 *personal protective equipment* such as eye protection or respiratory equipment.

Only once all of the measures numbered 1 to 5 above have been tried and found ineffective in controlling the risks to a reasonably practical level, must point 6, personal protective equipment (PPE), be used.

Audit

To conduct a systematic review to make sure standards and management systems are being followed.

Assessment criteria

3.3 Describe the procedures that should be taken to remove or minimise risks before deciding PPE is needed.

ACTIVITY

Identify three possible risks in using a ladder as access equipment when fixing a PIR luminaire 4 m high.

Assessment criteria

3.4 State the purpose of PPE.

Personal protective equipment (PPE)

All equipment, including clothing for weather protection, worn or held by a person at work, which protects that person from risks to health and safety.

PERSONAL PROTECTIVE EQUIPMENT AND ITS USE

Virtually all **personal protective equipment (PPE)** is covered by the Personal Protective Equipment at Work Regulations 1992 (PPE Regulations). The exception is respiratory equipment, which is covered by specific regulations relating to specific substances (lead, gases, substances hazardous to health, etc).

PPE is defined in the PPE Regulations as 'all equipment (including clothing affording protection against the weather) which is intended to be worn or held by a person at work and which protects them against one or more risks to his health or safety'. Such equipment includes safety helmets, gloves, eye protection, high-visibility clothing, safety footwear and safety harnesses. Employers are responsible for providing, replacing and paying for PPE.

Hearing protection and respiratory protective equipment provided for most work situations are not covered by these regulations because other regulations apply to them. However, these items need to be compatible with any other PPE provided.

PPE should be used only when all other measures are inadequate to control exposure. It protects only the wearer while it is being worn and, if it fails, PPE offers no protection at all. The provision of PPE is only one part of the protection package; training, selection of the correct equipment in all work situations, good supervision, monitoring and supervision of its use. All play a part in the success of PPE as a control measure.

Assessment criteria

3.5 Specify the appropriate protective clothing and equipment that is required for identified work tasks.

PPE FOR DIFFERENT TYPES OF PROTECTION

Protection for the eyes

Hazards – chemical or metal splash, dust, projectiles, gas and vapour, radiation.

PPE options – safety spectacles, goggles, face shields, visors.

Different types of eye protection

Protection for the head

Hazards – impact from falling or flying objects, risk of head bumping or hair getting caught.

PPE options – a range of helmets and bump caps.

A safety helmet and bump cap

Protection for breathing

Hazards – dust, vapour, gas, oxygen-deficient atmospheres.

PPE options – disposable filtering face-piece or respirator, half or full-face respirators, air-fed helmets, breathing apparatus.

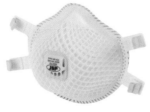

Examples of protective breathing equipment

Protection for hands and arms

Hazards – abrasion, temperature extremes, cuts and punctures, impact, chemicals, electric shock, skin infection, disease or contamination.

PPE options – gloves, gauntlets, mitts, wrist cuffs, armlets.

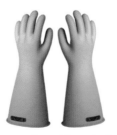

Riggers' gloves – for heavy duty/manual handling work, PVC gloves – for acids and oil, IEC 60903 gloves – made from insulating material, for live electrical work

ACTIVITY

At what stages might gloves be required during the electrical installation process?

Protection for the body

Hazards – temperature extremes, adverse weather, chemical or metal splash, spray from pressure leaks or spray guns, impact or penetration, contaminated dust, excessive wear or entanglement of own clothing.

PPE options – conventional or disposable overalls, boiler suits, specialist protective clothing such as chain-mail aprons, flame-retardant or high-visibility clothing.

Protection for feet and legs

Hazards – spillages underfoot, electrostatic build-up, slipping, cuts and punctures, falling objects, metal and chemical splash, abrasion.

PPE options – safety boots and shoes with protective toecaps and penetration-resistant mid-sole, gaiters, leggings, spats.

It is important that employees know why PPE is needed and are trained to use it correctly, as otherwise it is unlikely to protect as required. The following points should be considered.

- Does it fit correctly?
- How does the wearer feel?
- Is it comfortable?
- Do all items of PPE work well together?
- Does PPE interfere with the job being done?
- Does PPE introduce another health risk, for example, of overheating or getting caught up in machinery?
- If PPE needs maintenance or cleaning, how is it done?

When employees find PPE comfortable they are far more likely to wear it.

THE REQUIREMENT FOR FIRST-AID FACILITIES

In the event of injury or sudden illness, failure to provide first aid could result in a casualty's death. Employers must ensure that an employee who is injured or taken ill at work receives immediate attention.

The Health and Safety (First-Aid) Regulations 1981 require employers to provide adequate and appropriate equipment, facilities and personnel to ensure their employees receive immediate attention if they are injured or taken ill at work. These Regulations apply to all workplaces, including those with fewer than five employees and to the self-employed.

What is 'adequate and appropriate' will depend on the circumstances in the workplace and employers should carry out an assessment of first-aid needs to determine what to provide.

There are a number of factors to be considered:

- number of employees
- type of work being undertaken
- unusual hazards
- 'off-site' workers
- out of hours or shift patterns
- nearest emergency services.

This will determine whether trained first-aiders are needed, what should be included in a first-aid box and whether or not a first-aid room is required.

THE FIRST-AID BOX

There is no standard list of items that need to be contained in a first-aid box and the contents depend on what the employer or the 'person in control of the site' assesses the needs to be. However, tablets or medicines should not be kept in first-aid boxes. It is sensible to have an 'appointed person' to look after the first-aid equipment and replenish when required.

Emergencies can occur at any time and there may well be an immediate need for items without having to search for them. Items in first-aid boxes should not be taken for other uses. Items are needed for particular reasons and therefore misuse will decrease supplies.

Assessment criteria

3.7 Explain why it is important not to misuse first-aid equipment/supplies and to replace first-aid supplies once used.

AIM FOR BEST PRACTICE

Despite the range of statutory legislation and guidance that is available to make the workplace safe, accidents and injuries continue to happen. It is a misconception that death and serious injury only occur in what could be termed dangerous occupations such as scaffold erection, steeplejack work, mining, tunnelling, off-shore oil rigs, construction, fishing and fire fighting. Accidents can happen in any work activity and if some basic principles are observed most of these accidents can be prevented.

Most accidents are caused either by human error or environmental conditions.

Human error can be described as 'the failure of a worker to perform an assigned task correctly'.

Assessment criteria

3.8 Describe safe practices and procedures in the working environment.

ACTIVITY

Can you identify two actions which could be described as 'behaving foolishly'?

This error can occur at various stages of a work process — in the design, during manufacturing, during operation and during maintenance — and this can be for a number of reasons:

- lack of training
- carelessness
- behaving foolishly
- lack of attention
- tiredness
- lack of competence to carry out task
- drink or drugs.

Conditions termed 'environmental' are normally attributed to damaged or faulty tools and equipment, unguarded or faulty machinery, poor housekeeping (untidy workplace, incorrect storage of materials, trailing leads), poor workplace lighting or poor ventilation.

The most common categories of risk and causes of accidents at work when carrying out electrical installation are:

- fire
- electricity
- manual handling
- slips, trips and falls
- mechanical handling
- storage of goods and materials
- use of hand tools, including portable electric hand tools.

Construction sites and building works are hazardous environments in which to undertake work because of the temporary nature of the construction process, with the whole environment changing on a regular basis. So even when a risk assessment has been undertaken and written safe systems of work, such as a permit-to-work procedure, have been introduced, it is still a sensible precaution for an on-site safety assessment to be made by the operative to determine if any additional safety precautions are required before work starts.

While at work, employees are required to take reasonable care for the health and safety of themselves and others, and in order to discharge this responsibility it will be necessary to make the work area safe both before work starts and for the duration of the work. The following is an example of actions that will contribute to a safe working procedure:

- Provide demarcation of the safe zone of work by indicating the part of the system or items of electrical equipment on which it is safe to work, by using warning notices or other means.

- Provide effective control of the work area by restricting access and the use of barriers or screens. This can help in situations where the work is in such close proximity to live conductors that danger may arise.

- The system must be isolated from all points of supply, including auxiliary circuits, and dual or alternative supplies, such as uninterruptable power supplies (UPS) or a stand-by generator (eg on automatic start-up).

- Each point of isolation must be secured, for example by safety locking, removal of fuses, etc.

- If deemed necessary, a safety document should be issued, eg a permit-to-work (PTW).

- Tools should be in good condition, used correctly and maintained in a serviceable condition, ie cutting tools sharpened before use, screwdrivers maintained with square blades.

- Tools and materials should be stored safely when not in use. Suitable storage facilities should be provided but storage should never block work areas or evacuation routes.

ACTIVITY

When should tools and instruments be inspected for damage?

Items for promoting a safe working environment

Consideration should be given to the following to promote a safe working environment.

Tools

Where mains voltage (230 V) electrical power tools and extension leads are used, the risk of injury is high if equipment, tools or leads are damaged or there is a fault. Equipment should be visually checked for damage before use and be subjected to regular in-service inspections (EAW Regulation 4(4)) and taken out of service if damaged. A residual current device (RCD) is a device which detects some, but not all, faults in the electrical system and rapidly switches off the supply. Cordless tools or those that operate from a 110 V centre tapped to earth (CTE) supply will reduce the risk of injury.

ASSESSMENT GUIDANCE

Battery tools offer a convenient method of reducing shock risk and problems with trailing leads. 110 V equipment is supplied via a 230 V to 110 V transformer with the secondary centre tap connected to earth (CTE). This reduces the maximum shock voltage to earth to 55 V. This is known as a reduced low voltage system.

Cordless tools

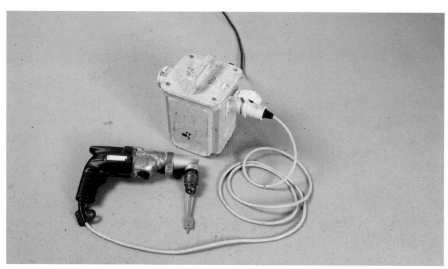

A 110 V drill and transformer

RCDs must be properly installed and enclosed, checked daily, treated with care, kept free of moisture and dirt, and protected against vibration and mechanical damage.

50% of all trip accidents are caused by bad housekeeping. Tools and equipment left lying about the workplace contribute to slip, trips and falls and if not returned to their storage boxes/containers, can also be stolen. Materials that are used in the work process, for example, cable, wiring accessories, luminaires, conduit and trunking etc, should all be stored correctly to avoid them becoming a tripping hazard.

Lighting

The lighting of switch rooms, etc needs to be considered from the work as well as the operational point of view. A general lighting level that is adequate for operating electrical equipment may be insufficient for work on the equipment, as the lights may cast shadows in some areas. Provision must be made where necessary for the use of portable inspection lamps (operated from a reduced voltage source).

Regulation 21 of PUWER places a general duty on employers to ensure that there is suitable and sufficient lighting, taking account of the operations to be carried out. This complements a similar requirement contained in Regulation 8 of the Workplace (Health, Safety and Welfare) Regulations 1992 (WHSWR).

Access is likely to be required at times of failure of the normal supply and arrangements should therefore exist for some form of emergency lighting. Where there is no permanent emergency lighting installed, adequate illumination should be available by other means, such as portable hand lamps.

Regulation 8 of the WHSWR requires that suitable and sufficient emergency lighting be provided in any circumstances where persons at work are specially exposed to danger in the event of failure of artificial lighting, as could be the case with 'live working'.

HSE Guidance Booklet HSG38 gives guidance on lighting at work generally and for different work activities. The Chartered Institution of Building Services Engineers (CIBSE) indoor lighting guide recommends a minimum lighting level of 200 lux in electrical switch rooms.

Chemicals

The European chemicals industry manufactures and uses a large number of different chemical products. 90–95% of all chemicals on the European market are preparations, ie mixtures of chemical substances. They include industrial chemicals, such as solvents and coatings; petrochemicals, such as fuels and lubricants; agricultural chemicals, such as pesticides; consumer products, such as detergents and disinfectants; and many others. Whereas the majority of these chemicals are of low concern for human health or the environment, some of them do have properties which are hazardous to human health and/or the environment.

There are two key pieces of EU legislation aimed at achieving a high level of protection of human health and the environment from chemicals, by requiring the producers of chemicals to:

■ identify the intrinsic hazards of the chemicals they manufacture or import

■ label the chemicals

■ package them in a safe way.

These requirements are dealt with in more detail in 'The requirement for hazard identification' on page 51.

'The requirement for hazard identification' on page 51.

WORKING AT HEIGHT

Working at height remains one of the biggest causes of fatalities and major injuries within the construction industry, with almost 50% of fatalities resulting from falls from ladders, stepladders and through fragile roofs. Work at height means work in any place, including at or below ground level (eg in underground workings), where a person could fall a distance liable to cause injury.

The Work at Height Regulations 2005 require duty holders to ensure that:

■ all work at height is properly planned and organised

■ those involved in work at height are competent

■ the risks from work at height are assessed and appropriate work equipment is selected and used

■ the risks of working on or near fragile surfaces are properly managed

■ the equipment used for work at height is properly inspected and maintained.

ACTIVITY

Identify two chemicals or substances which are commonly used by electricians.

ASSESSMENT GUIDANCE

Remember that standing on the floor of a six-storey building is not really working at height, but standing on a stepladder on the ground floor is.

For information and guidance regarding mobile access towers, the Prefabricated Access Suppliers' and Manufacturers' Association (PASMA), is the recognised focus and authority. PASMA provides training and governs the industry.

ACCESS EQUIPMENT FOR DIFFERENT TYPES OF WORK

Many different types of access equipment are used in the building services industry, such as mobile elevated work platforms (MEWPs), ladders, mobile tower scaffolds, tube and fitting scaffolding, personal suspension equipment (harnesses).

MEWPs

There is a wide range of MEWPs (vertical scissor lift, self-propelled boom, vehicle-mounted boom and trailer-mounted boom) and if any of these are to be used it is important to consider:

- the height from the ground
- whether the MEWP is appropriate for the job

A vertical scissor lift

- the ground conditions
- training of operators
- overhead hazards such as trees, steelwork or overhead cables
- the use of a restraint or fall arrest system
- closeness to passing traffic.

Ladders

It is recommended that ladders are only used for low-risk, short-duration work (between 15 and 30 minutes depending upon the task). Common causes of falls from ladders are:

- overreaching – maintain three points of contact with the ladder
- slipping from the ladder – keep the rungs clean, wear non-slip footwear, maintain three points of contact with the ladder, make sure the rungs are horizontal
- the ladder slips from its position – position the ladder on a firm level surface, secure the ladder top and bottom, check the ladder daily
- the ladder breaks – position the ladder properly using the 1:4 rule (four units up for every one unit out), do not exceed the maximum weight limit of the ladder, only carry light materials or tools.

The correct angle for a ladder is 1:4

Stepladders

Many of the common causes of falls from ladders can also be applied to stepladders. If stepladders are used, ensure that:

- they are suitable, in good condition and inspected before use
- they are sited on stable ground
- they are face onto the work activity
- knees are never above the top tread of the stepladder
- the stepladder is open to the maximum
- wooden stepladders (or ladders) are not painted, as this may hide defects.

Scaffolding

Some work activities, such as painting, roof work, window replacement or brickwork, are almost certainly more conveniently undertaken from a fixed external scaffold. Tube and fitting scaffolding must only be erected by competent people who have attended recognised training courses. Any alterations to the scaffolding must also be carried out by a competent person. Regular inspections of the scaffold must be made and recorded.

Correct use of stepladders

SAFETY CHECKS AND SAFE ERECTION METHODS FOR ACCESS EQUIPMENT

The Work at Height Regulations 2005 require that all scaffolding and equipment that is used for working at height, where a person could fall 2 metres or more, is inspected on a regular basis, using both formal and pre-use inspections to ensure that it is fit for use.

A marking, coding or tagging system is a good method of indicating when the next formal inspection is due. However, regular pre-use checks must take place as well as formal inspections.

The following safety checks must be carried out.

Tags to record that a ladder inspection has taken place

MEWPs

These must only be operated by trained and competent persons, who must also be competent to carry out the following pre-use checks.

- The ground conditions must be suitable for the MEWP, with no risk of the MEWP becoming unstable or overturning.
- Check for overhead hazards such as trees, steelwork and overhead cables.
- Guard rails and toe boards must be in place.
- Outriggers should be fully extended and locked in position.
- The tyres must be properly inflated and the wheels immobilised.
- Check the controls to make sure they work as expected.
- Check the fluid and/or battery charge levels.
- Check that the descent alarm and horn are working.
- Check that the emergency or ground controls are working properly.

Tube and fitting scaffolds

This equipment must only be erected by competent people who have attended recognised training courses. Any alterations to the scaffolding must also be carried out by a competent person. However, users should undertake the following fundamental checks to prevent accidents.

- Toe boards, guard rails and intermediate rails must be in place to prevent people and materials from falling.
- The scaffold must be on a stable surface and the uprights must be fitted with base plates and sole plates.
- Safe access and exit (ladders) must be in place and secured.

Ladders and step ladders

Users must check that:

- ladders and stepladders are of the right classification (trade/industrial)

- the styles, rungs or steps are in good condition
- the feet are not missing, loose, damaged or worn
- rivets are in place and secure
- the locking bars are not bent or buckled.

Prefabricated mobile scaffold towers

Mobile scaffold towers are a convenient means of undertaking repetitive tasks in the building services industry. The erection and dismantling must only be undertaken by a competent person. The following pre-use checks will ensure safe use.

- The maximum height-to-base ratios must not be exceeded.
- Diagonal bracing and stabilisers must not be damaged or bent.
- The brace claws must work properly.
- Internal access ladders must be in place.
- Wheels must be locked when work is in progress.
- The working platform must be boarded, with guard rails and toe boards fitted.
- The towers must be tied in windy conditions.
- The platform trap door must be operating correctly.
- Rivets must be checked visually to ensure they are in place and not damaged.

ASSESSMENT GUIDANCE

You must always dismount a mobile scaffold before it is moved or adjusted.

SAFETY SIGNS

The Health and Safety (Safety Signs and Signals) Regulations 1996 require employers to ensure that safety signs are provided (or are in place) and maintained in circumstances where risks to health and safety have not been avoided by other means, for example, safe systems of work. The range of safety signs are shown on the following pages.

Prohibition signs

These signs indicate an activity that must not be done. They are circular white signs with a red border and red cross bar.

No access for unauthorised persons

No smoking in this building

In the event of fire do not use this lift

Do not drink

Some prohibition signs

Warning signs

These provide safety information and/or give warning of a hazard or danger. They are triangular yellow signs with a black border and symbol.

Some warning signs

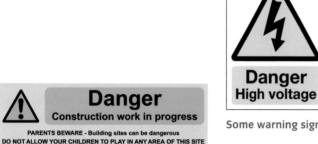

A typical construction site safety sign

Mandatory signs

These signs give instructions that must be obeyed. They are circular blue signs with a white symbol.

Some mandatory signs

Advisory or safe condition signs

These give information about safety provision. They are square or rectangular signs with a white symbol.

Some safe condition signs

Understand the requirements for identifying and dealing with hazards in the work environment

THE REQUIREMENT FOR HAZARD IDENTIFICATION

A hazardous substance is defined as something that can cause ill-health to people at work.

This may include the substances used directly in the work process, such as glue, paints, thinners, solvents and cleaning materials, and those produced by different work activities, for example, welding fume. Health hazards are ever-present during building services activities due to the nature of the activities and may also include other hazards such as vibration, dust (possibly including asbestos), cement and solvents.

There are two key pieces of EU legislation aimed at achieving a high level of protection of human health and the environment from chemicals, by requiring the producers of chemicals:

- to identify the intrinsic hazards of the chemicals they manufacture or import (ie to 'classify' chemicals according to their dangers such as flammability, toxicity, carcinogenicity)
- to label these chemicals according to strict rules (ie warnings about the dangers and safety advice)
- to package them safely.

The Dangerous Substances Directive sets out harmonised EU rules for classification, packaging and labelling of dangerous chemical substances and the Dangerous Preparations Directive extends these rules to dangerous preparations.

These directives are being introduced within the EU as the Regulation on Classification, Labelling and Packaging of Substances and Mixtures (known as the CLP Regulation) and the present UK regulations (CHIP) will be gradually replaced by the CLP Regulation between 2010 and 2015.

The Dangerous Substances Directive and the Dangerous Preparations Directive were implemented in Great Britain as the Chemicals (Hazard Information and Packaging for Supply) Regulations 2009: (CHIP). The CHIP regulations have been amended to ensure that national law is

Assessment criteria

4.1 Identify warning signs for the seven main groups of hazardous substance, as defined by The Chemical (Hazard Information and Packaging for Supply) Regulations (CHIP).

SmartScreen Unit 301
Handout 28 and Worksheet 28

aligned with the transitional arrangements in the CLP Regulation, and to include the necessary enforcement provisions where chemical suppliers choose to apply the CLP provisions as an alternative to CHIP ahead of the mandatory compliance dates.

CHIP requires the supplier of a dangerous chemical to:

- identify the hazards (dangers) of the chemical — this is known as 'classification'
- give information about the hazards to their customers — suppliers usually provide this information on the package itself (eg a label)
- package the chemical safely.

The seven main groups of hazardous substances, as defined in CHIP, are explained below.

Flammable

(F+)   Extremely flammable. Chemicals that have an extremely low flash point and boiling point, and gases that catch fire in contact with air

(F)   Highly flammable. Chemicals that may catch fire in contact with air, only need brief contact with an ignition source, have a very low flash point or evolve highly flammable gases in contact with water.

Explosive

(E)   Chemicals that explode.

Oxidising

(O)   Chemicals that react exothermically with other chemicals.

Irritant

(X)   Chemicals that may cause inflammation to the skin or other mucous membranes.

An irritant is a non-corrosive substance that can cause skin or lung inflammation after repeated exposure. White spirit, bleach and glues are common irritants.

Toxic

(T+)   Chemicals that at very low levels cause damage to health.

(T)   Chemicals that at low levels cause damage to health.

A toxic substance is one that will impede or prevent the function of one or more of the organs in the body (kidneys, liver, heart). Lead, mercury and pesticides are toxic substances.

Corrosive

(C)   Chemicals that may destroy living tissue on contact.

Dangerous to the environment

(N)   Chemicals that may present an immediate or delayed danger to one or more components of the environment.

SYMBOLS FOR HAZARDOUS SUBSTANCES

Hazardous substances are given a classification according to the severity and type of hazard they may present to people in the workplace. However, all over the world there are different laws on how to identify the hazardous properties of chemicals. The United Nations has therefore created the Globally Harmonised System of Classification and Labelling of Chemicals (GHS). The aim of the GHS is to have, worldwide, the same:

- criteria for classifying chemicals according to their health, environmental and physical hazards
- hazard communication requirements for labelling and safety data sheets.

The GHS is not a formal treaty, but is a non-legally binding international agreement. In Great Britain the existing legislation is the Chemicals (Hazard Information and Packaging for Supply) Regulations 2009 (CHIP), to be superseded by the European Classification, Labelling and Packaging of Substances and Mixtures Regulations by 2015.

KEY POINT

You do not need to know all of the detail in the GHS, but it is important that you are aware of it.

SmartScreen Unit 301

Handout 22 and Worksheet 22

Examples of the GHS labelling system are shown below:

**The hazard signs shown indicate substances that are
(top row, left to right) flammable, explosive, oxidising, gas under pressure, toxic and
(bottom row, left to right) long-term health hazards (causes of cancer), corrosive,
irritant (requires caution), dangerous to the environment**

Toxic

Poisonous.

COMMON HAZARDOUS SUBSTANCES

Hazardous substances at work may include those used directly in the work process, such as glue, paints, thinners, solvents and cleaning materials, and those produced by different work activities, such as welding fumes. Health hazards are always present during building services activities due to the nature of the activities and may include hazards, such as vibration, dust (possibly including asbestos) as well as cement and solvents.

Hazardous substance

Something that can cause ill health to people.

The Control of Substances Hazardous to Health (COSHH) Regulations 2002 provide a framework that helps employers assess risk and monitor effective controls. A COSHH assessment is essential before work starts and should be updated as new substances are introduced.

Hazardous substances include:

- any substance that gives off fumes that may cause headaches or respiratory irritation

- acids that cause skin burns or respiratory irritation (eg battery acid, metal-cleaning materials)

ACTIVITY

Think of some common electrical components that could contain substances hazardous to health.

- solvents, for example, for PVC tubes and fittings, that cause skin and respiratory irritation

- man-made fibres that cause eye or skin irritation (eg thermal insulation, optical fibres)

- cement and wood dust that may cause eye irritation and respiratory irritation

- fumes and gases that cause respiratory irritation (eg soldering, brazing and welding fumes, or overheating/burning PVC)

- asbestos.

The COSHH Regulations require employers to assess risk and ensure the prevention or adequate control of exposure to hazardous substances by measures other than the provision of personal protective equipment (PPE) 'so far as is reasonable practicable'.

PRECAUTIONS TO BE TAKEN WITH HAZARDOUS SUBSTANCES

There is an acknowledged 'hierarchy of control' list of measures designed to control risks. These are considered in order of importance, effectiveness or priority. The measures are listed as follows:

- eliminate the risk by designing out or changing the process
- reduce the risk by substituting less hazardous substances
- isolate the risk using enclosures, barriers or by moving workers away
- control the risk by the introduction of guarding or local exhaust systems
- management control such as safe systems of work, training, etc
- PPE such as eye protection or respiratory equipment.

UNDERSTANDING HAZARDS

The HSW Act requires employers to provide health and safety systems and procedures in the workplace. Directors and managers of companies employing more than five employees can be held personally responsible for health and safety failures. The Management of Health and Safety at Work Regulations 1999 detail and encourage a more systematic and better organised approach to dealing with health and safety in all workplaces.

The control of risks is essential to the provision and maintenance of a safe and healthy workplace and ensures compliance with MHSWR Regulation 3. Risk assessment starts with the need for hazard identification.

A hazard is usually defined as anything with the potential to cause harm (chemicals, a fault on electrical equipment, working at height). Risk is the chance (large or small) of harm actually being done when things go wrong (eg risk of electric shock from faulty equipment).

These are usually evaluated in terms of:

- the likelihood of occurrence, ie is a hazard going to occur
- the severity of outcome, ie how serious the resulting injury will be.

ASSESSMENT GUIDANCE

Make sure you can recognise a hazard. Be able to explain what you can do to reduce the risk.

ACTIVITY

How would you safely dispose of discharge lamps containing mercury?

ACTIVITY

What hazardous material may be found in a smoke detector?

Assessment criteria

4.2 Define what is meant by the term 'hazard' in relation to health and safety legislation in the workplace.

All organisations must have the freedom to prepare systems of work that match the risk potential of their particular work activity and which are practical in their application.

Electrotechnical work operations are no different to other work disciplines in this respect and employers must examine the work activities and the workplace and introduce systems of work through a process of 'risk assessment' that takes account of the risks and the application of the health and safety legislation designed to manage those risks.

The work operations that may be undertaken in the electrotechnical field will all have particular hazards and these and the relevant health and safety legislation are considered in more detail in the following section.

UNDERSTANDING HAZARDS DURING ELECTROTECHNICAL WORK

Assessment criteria

4.3 Identify specific hazards associated with the installation and maintenance of electrotechnical systems and equipment.

As has been discussed previously, the hazards associated with the installation and maintenance of electrotechnical systems and equipment are very wide ranging and will be influenced by the environment in which the work is being undertaken. The specific hazards, however, can be listed as follows.

Electrical injuries can be caused by a wide range of voltages, and are dependent upon individual circumstances, but the risk of injury is generally greater with higher voltages. Alternating current (a.c.) and direct current (d.c.) electrical supplies can cause a range of injuries including:

- electric shock
- electrical burns
- loss of muscle control
- fires arising from electrical causes
- arcing and explosion.

Electric shock

ASSESSMENT GUIDANCE

It is a very good idea to undergo training about general first aid but especially electric shock treatment. You never know when it will be needed.

Electric shocks may arise either by direct contact with a live part or indirectly by contact with an exposed conductive part (eg a metal equipment case) that has become live as a result of a fault condition. Faults can arise from a variety of sources:

- broken equipment case exposing internal bare live connections
- cracked equipment case causing 'tracking' from internal live parts to the external surface
- damaged supply cord insulation, exposing bare live conductors
- broken plug, exposing bare live connections.

The magnitude (size) and duration of the shock current are the two most significant factors determining the severity of an electric shock. The magnitude of the shock current will depend on the contact voltage and impedance (electrical resistance) of the shock path. A possible shock path always exists through ground contact (eg hand-to-feet). In this case the shock path impedance is the body impedance plus any external impedance. A more dangerous situation is a hand-to-hand shock path when one hand is in contact with an exposed conductive part (eg an earthed metal equipment case), while the other simultaneously touches a live part. In this case the current will be limited only by the body impedance.

As the voltage increases, so the body impedance decreases, which increases the shock current. When the voltage decreases, the body impedance increases, which reduces the shock current. This has important implications concerning the voltage levels that are used in work situations and highlights the advantage of working with reduced low voltage (110 V) systems or battery-operated hand tools.

At 230 V, the average person has a body impedance of approximately 1 300 Ω. At mains voltage and frequency (230 V–50 Hz), currents as low as 50 milliamps (0.05 A) can prove fatal, particularly if flowing through the body for a few seconds.

ACTIVITY

Battery-powered equipment is much safer and more convenient than mains equipment, but generally lacks the constant power available with 110 V equipment. Are there any circumstances when 230 V equipment can be used on site?

Electrical burn

Burns may arise due to:

- the passage of shock current through the body, particularly if at high voltage
- exposure to high-frequency radiation (eg from radio transmission antennas).

ACTIVITY

Static electricity can also cause electric shock, though it is more of a nuisance than a danger. How is static electricity produced?

Loss of muscle control

People who experience an electric shock often get painful muscle spasms that can be strong enough to break bones or dislocate joints. This loss of muscle control often means the person cannot 'let go' or escape the electric shock. The person may fall if they are working at height or be thrown into nearby machinery or structures.

Fire

Electricity is believed to be a factor in the cause of over 30 000 fires in domestic and commercial premises in Britain each year. One of the principal causes of such fires is wiring with defects such as insulation failure, the overloading of conductors, lack of electrical protection or poor connections.

Arcing

Arcing frequently occurs due to short-circuit flashover accidentally caused while working on live equipment (either intentionally or unintentionally). Arcing generates UV radiation, causing severe sunburn. Molten metal particles are also likely to be deposited on exposed skin surfaces. Arcing or flashover can also cause temporary blindness known as 'arc-eye' or, in some cases, permanent blindness.

Explosion

There are two main electrical causes of explosion: short circuit due to an equipment fault, and ignition of flammable vapours or liquids caused by sparks or high surface temperatures.

Assessment criteria

4.4 Describe situations which can constitute a hazard in the workplace.

COMMON SITE HAZARDS

Temporary electrical supplies

The HSW Act and the EAW Regulations do not differentiate between permanent and temporary electrical supplies and both require the provision of safe plant and systems of work whatever the situation.

The **duty holder** or responsible person and the equipment/system designer in particular, will need to ensure that all system components selected are appropriate to the intended use, and rated accordingly. Two main factors have to be considered:

(i) the electrical system to which the equipment is to be connected. Both normal and possible abnormal fault conditions have to be assessed

(ii) the equipment specification and suitability for the intended use and environment.

Cables must be able to carry fault currents safely for the time taken by the protective device to operate to disconnect the fault. Correct cabling sizing also needs to be considered when temporary lighting is used for remote working. Long lighting leads or extension leads could affect the earth fault loop impedance and the ability to provide adequate electric shock protection. Consideration may need to be given to providing individual RCD protection or separate earth leads for external lighting circuits.

It is preferable for multi-core cables to be used for temporary distribution systems.

Cables need to be flexible and of suitable size and mechanical strength for their intended duty. Cables for indoor use should be PVC or rubber sheathed as specified in *BS 6500: Electric Cables. Flexible cords rated up to 300/500 V for use with appliances and equipment intended for domestic, office and similar environments* (or equivalent), and cables for outdoor use should be rubber insulated and sheathed as specified in BS 7919. Cables of different ratings should only be connected through a distribution unit providing suitable overcurrent protection for all cables.

Switchgear and fuse gear must be selected so that their rating is appropriate for the prospective fault level at their points of installation. The duty imposed by EAW Regulation 5 in respect of the strength and capability of equipment not being exceeded, relates only to where danger may occur. In some cases, the normal load rating of equipment may be safely exceeded for a short duration where agreed by the manufacturer, eg on a cyclic loading basis (transformer).

EAW Regulation 6 requires that electrical equipment which may reasonably foreseeably be exposed to

- mechanical damage
- the effects of the weather, natural hazards, temperature or pressure
- the effects of wet, dirty, dusty or corrosive conditions
- any flammable or explosive substance, including dusts, vapours or gases

shall be of such construction or as necessary protected, to prevent, so far as is reasonably practicable, danger arising from such exposure.

The level of duty is 'so far as is reasonably practicable'.

It is necessary to take action in respect of those adverse conditions that are 'reasonably foreseeable'. The onus is on the responsible person/designer to specify the appropriate equipment for the installation environment.

Common examples of conditions which require close attention are:

- surface cabling running over the ground – this needs to be protected from the risk of mechanical damage
- equipment intended for outdoor use (and, where equipment is used indoors, the building must stay weatherproof)
- equipment for use in flammable or explosive atmospheres.

There is a wide range of guidance material, standards and codes of practice available:

- BS EN 60529:1992 contains the IP code (international protection), which is used to indicate the degree of protection afforded against the ingress of dirt, dust, moisture etc.

- BS EN 60079 covers electrical apparatus for use in potentially explosive atmospheres. However, there are proposals to introduce new EC Directives, known as the ATEX Directives, to give a more controlled approach to equipment certification throughout the EC.

- In Europe, EN 50102 'Degrees of protection provided by enclosures for electrical equipment against external mechanical impacts' (IK Code) has been introduced. In general, where given, the degree of protection will apply to a complete enclosure.

Manual handling techniques

Manual handling is one of the most common causes of injury at work and causes over a third of all workplace injuries. Manual handling injuries can occur almost anywhere in the workplace and heavy manual labour, awkward postures and previous or existing injury can increase the risk. Work-related manual handling injuries can have serious implications for both the employer and the person who has been injured.

The introduction of the Manual Handling Operations Regulations 1992 saw a change from reliance on safe lifting techniques to an analysis, using risk assessment, of the need for manual handling. The regulations established a clear hierarchy of manual handling measures:

- avoid manual handling operations so as far as is reasonably practicable by re-engineering the task to avoid moving the load or by mechanising the operation

- if manual handling cannot be avoided, a risk assessment should be made

- reduce the risk of injury so as far as is reasonably practicable either by the use of mechanical aids or by making improvements to the task (for example by using two persons), the load and the environment.

How to manually handle loads using mechanical lifting aids

Even if mechanical handling methods are used to handle and transport equipment or materials, hazards may still be present in the four elements that make up the mechanical handling: the handling equipment, the load, the workplace and the human element.

- The handling equipment – mechanical handling equipment must be suitable for the task, well maintained and inspected on a regular basis.

- The load – the load needs to be prepared in such a way as to minimise accidents, taking into account such things as security of the load, flammable materials and stability of the load

Manual handling

The movement of items by lifting, lowering, carrying, pushing or pulling by human effort alone.

ACTIVITY

What should you check before moving an object from one place to another by manual handling?

ASSESSMENT GUIDANCE

When you are manually handling a load, always make sure the route you will be taking is clear and all obstacles are removed. Never try to lift more than you can handle. Use safe handling techniques. Get someone to help you or use mechanical assistance to move the object.

- The workplace – if possible, the workplace should be designed to keep the workforce and the load apart

- The human element – employees who use the equipment must be properly trained.

Housekeeping

This is one of the most important single items influencing safety within the workplace. Poor housekeeping not only causes an increase in the risk of fire, slips, trips and falls, but may also expose members of the public to risks created during building services engineers' work, such as fault diagnosis and maintenance, as these activities are normally carried out in occupied buildings. The following are some examples of good housekeeping

- Stairways, passages and gangways should be kept free from materials, electrical power leads and obstructions of every kind.

- Materials and equipment should be stored tidily so as not to cause obstruction and should be kept away from the edges of buildings, roofs, ladder access, stairways, floor openings and rising shafts.

- Tools should not be left where they may cause tripping or other hazards. Tools not in use should be placed in a tool belt or tool bag and should be collected and stored in an appropriate container at the end of each working day.

- Working areas should be kept clean and tidy. Scrap and rubbish must be removed regularly and placed in proper containers or disposal areas.

- Rooms and site accommodation should be kept clean. Soiled clothes, scraps of food, etc should not be allowed to accumulate, especially around hot pipes or electric heaters.

- The spillage of oil or other substances must be contained and cleaned up immediately.

Tools should be kept neatly in a tool belt or bag

A badly managed site will inevitably lead to accidents

■ All flammable liquids, liquified petroleum gas (LPG) and gas cylinders must be stored properly.

Gas bottles should be stored properly

Slips, trips and falls

These are the most common hazards to people as they walk around the workplace. They make up over a third of all major workplace injuries. Over 10 000 workers suffered serious injury because of a slip or trip in 2011. Listed below are the main factors that can play a part in contributing to a slip- or trip-free environment.

Flooring

■ The workplace floor must be suitable for the type of work activity that will be taking place on it.

■ Floors must be cleaned correctly to ensure they do not become slippery, or be of a non-slip type that it keeps its slip-resistance properties.

■ Flooring must be fitted correctly to ensure that there are no trip hazards and any non-slip coatings must be correctly applied.

■ Floors must be maintained in good order to ensure that there are no trip hazards such as holes, uneven surfaces, curled-up carpet edges, or raised telephone or electrical sockets.

■ Ramps, raised platforms and other changes of level should be avoided. If they cannot be avoided, they must be highlighted.

Stairs

Stairs should have:

■ high-visibility, non-slip, square nosings on the step edges
■ a suitable handrail
■ steps of equal height
■ steps of equal width.

ACTIVITY

Use the internet to find suppliers of signs and barriers suitable for preventing access to the work area.

Contamination

Most floors only become slippery once they become contaminated. If **contamination** can be prevented, the slip risk can be reduced or eliminated.

Contamination of a floor can be classed as anything that ends up on a floor, including rainwater, oil, grease, cardboard, product wrapping, dust, etc. It can be a by-product of a poorly controlled work process or be due to bad weather conditions.

Contamination

The introduction of a harmful substance to an area.

Cleaning

Cleaning is important in every workplace; nowhere is exempt. It is not just a subject for the cleaning team. Everyone's aim should be to keep their workspace clear and deal with contamination such as spillages as soon as they occur.

The process of cleaning can itself create slip and trip hazards, especially for those entering the area being cleaned, including those undertaking the cleaning. Smooth floors left damp by a mop are likely to be extremely slippery. Trailing wires from a vacuum cleaner or polishing machine can present an additional trip hazard.

As people often slip on floors that have been left wet after cleaning, access to smooth wet floors should be restricted by using barriers, locking doors or cleaning in sections. Signs and warning cones only warn of a hazard, they do not prevent people from entering the area. If the water on the floor is not visible, signs and cones are usually ignored.

Human factors

How people act and behave in their work environments can affect the risk of slips and trips. For example:

- having a positive attitude toward health and safety. Dealing with a spillage instead of waiting for someone else to deal with it, can reduce the risk of slip and trip accidents
- wearing the correct footwear can also make a difference
- lack of concentration and distractions, such as being in a hurry, carrying large objects or using a mobile phone, all increase the risk of an accident.

ACTIVITY

It is very easy to trip or slip, which can result in an injury. What action should you take if you find that a plasterer has left material in a passageway?

Environmental factors

Environmental issues can affect slips and trips. The following points give an indication of these issues.

- Too much light on a shiny floor can cause glare and stop people from seeing hazards, for example, on floors and stairs.
- Too little light will prevent people from seeing hazards on floors and stairs.

- Unfamiliar and loud noises may be distracting.
- Rainwater on smooth surfaces inside or outside a building may create a slip hazard.

Footwear

Footwear can play an important part in preventing slips and trips. The following highlights some relevant points.

- Footwear can perform differently in different situations, for example, footwear that performs well in wet conditions might not be suitable where there are food spillages.
- A good tread pattern on footwear is essential on fluid-contaminated surfaces.
- Sole tread patterns should not be allowed to become clogged with any waste or debris on the floor, as this makes them unsuitable for their purpose.
- Sole material type and hardness are key factors influencing safety.
- The choice of footwear should take into account factors such as comfort and durability, as well as obvious safety features such as protective toecaps and steel mid-soles.

Presence of dust and fumes

Electrotechnical work activities can present all sorts of dangers to the lungs, such as dust from insulation and ceiling tiles, mist from paint spraying, and fumes from brazing or soldering. The provision of proper ventilation is beneficial when working in almost every aspect of electrotechnical work. This can be supplemented with more specific safety measures, such as respirators and dust collection systems, when necessary, but overall ventilation is still the first choice. Without ventilation, other workers and visitors may be put at risk. Harmful particles will often remain in the air for a period long after the task has been completed and any PPE such as a respirator has been removed.

When a solid is heated to its melting point, it may release a fume, which is a solid particle suspended in the air. Welding and soldering are common practices which create fumes.

Sanding, grinding, and even handling powders such as plaster, can create dust. Like fumes, these are solid particles floating in the air. Though most dust is trapped by nasal hairs, some dust is so fine it can penetrate to the lungs. Dusts of this type are known as 'respirable' dusts, and are the most harmful. Asbestos dust and fibres are in this category. Some dusts are so fine they are invisible.

Tiny liquid droplets in the air are known as a mist. Mists can be created by spraying liquid, or by boiling it.

ASSESSMENT GUIDANCE

If you need to wash a floor to remove dust, remember to put up warning signs regarding the slippery floor.

Hazardous malfunctions of equipment

If a piece of equipment fails in its function, ie fails to do what it was supposed to do and as a result this failure has the potential to cause harm, then this would be defined as a hazardous malfunction.

The Provision and Use of Work Equipment Regulations 1998 (PUWER) require that equipment provided for use at work, including machinery, is:

- suitable for the intended use
- safe for use, maintained in a safe condition and inspected to ensure it is correctly installed and does not subsequently deteriorate
- used only by people who have received adequate information, instruction and training
- accompanied by suitable health and safety measures, such as protective devices and controls (emergency stop devices, adequate means of isolation and clear visible markings and warning devices).

Hazardous malfunctions are considered as dangerous occurrences under RIDDOR and must be reported in accordance with the RIDDOR reporting procedure.

WORKING AT HEIGHT

Working at height remains one of the biggest causes of fatalities and major injuries within the construction industry, with almost 50% of fatalities resulting from falls from ladders, stepladders and through fragile roofs. Work at height means work in any place, including at or below ground level (eg in underground workings), where a person could fall a distance liable to cause injury.

ACTIVITY

What would constitute working at height?

The Work at Height Regulations 2005 require duty holders to ensure that:

- all work at height is properly planned and organised
- those involved in work at height are competent
- the risks from work at height are assessed and appropriate work equipment is selected and used
- the risks of working on or near fragile surfaces are properly managed
- the equipment used for work at height is properly inspected and maintained.

For information and guidance regarding mobile access towers, the Prefabricated Access Suppliers' and Manufacturers' Association (PASMA), is the recognised focus and authority. PASMA provides training and governs the industry.

ASSESSMENT GUIDANCE

There is no shame in admitting you suffer from vertigo. Do not put yourself at risk. Use all required safety equipment.

ACCESS EQUIPMENT FOR DIFFERENT TYPES OF WORK

Many different types of access equipment are used in the building services industry, such as mobile elevated work platforms (MEWPs), ladders (including stepladders), mobile tower scaffolds, tube and fitting scaffolding, personal suspension equipment (harnesses).

MEWPs

There is a wide range of MEWPs (vertical scissor lift, self-propelled boom, vehicle-mounted boom and trailer-mounted boom) and these are considered in more detail in Learning outcome 3.8 on pages 46 and 48.

Ladders

It is recommended that ladders are only used for low-risk, short-duration work (between 15 and 30 minutes depending upon the task). Common causes of falls from ladders are:

- overreaching – maintain three points of contact with the ladder
- slipping from the ladder – keep the rungs clean, wear non-slip footwear, maintain three points of contact with the ladder, make sure the rungs are horizontal
- the ladder slips from its position – position the ladder on a firm level surface, secure the ladder top and bottom, check the ladder daily
- the ladder breaks – position the ladder properly using the 1:4 rule (four units up for every one unit out), do not exceed the maximum weight limit of the ladder, only carry light materials or tools.

Stepladders

Many of the common causes of falls from ladders can also be applied to stepladders. If stepladders are used, ensure that:

- they are suitable, in good condition and inspected before use
- they are sited on stable ground
- they are face onto the work activity
- knees are never above the top tread of the stepladder
- the stepladder is open to the maximum
- wooden stepladders (or ladders) are not painted, as this may hide defects
- they are of the right classification (trade/industrial)
- the styles, rungs or steps are in good condition
- the feet are not missing, loose, damaged or worn

- rivets are in place and secure
- the locking bars are not bent or buckled.

Scaffolding

Some work activities, such as painting, roof work, window replacement or brickwork, are almost certainly more conveniently undertaken from a fixed external scaffold.

Tube and fitting scaffolding must only be erected by competent people who have attended recognised training courses. Any alterations to the scaffolding must also be carried out by a competent person. Regular inspections of the scaffold must be made and recorded.

Tube and fitting scaffolds

This equipment must only be erected by competent people who have attended recognised training courses. Any alterations to the scaffolding must also be carried out by a competent person. However, users should undertake the following fundamental checks to prevent accidents.

- Toe boards, guard rails and intermediate rails must be in place to prevent people and materials from falling.
- The scaffold must be on a stable surface and the uprights must be fitted with base plates and sole plates.
- Safe access and exit (ladders) must be in place and secured.

Prefabricated mobile scaffold towers

Mobile scaffold towers are a convenient means of undertaking repetitive tasks in the building services industry. The erection and dismantling must only be undertaken by a competent person. The following pre-use checks will ensure safe use.

- The maximum height-to-base ratios must not be exceeded.
- Diagonal bracing and stabilisers must not be damaged or bent.
- The brace claws must work properly.
- Internal access ladders must be in place.
- Wheels must be locked when work is in progress.
- The working platform must be boarded, with guard rails and toe boards fitted.
- The towers must be tied in windy conditions.
- The platform trap door must be operating correctly.
- Rivets must be checked visually to ensure they are in place and not damaged.

Assessment criteria

4.5 Explain practices and procedures for addressing hazards in the workplace.

ADDRESSING HAZARDS AND RISK

As was discussed in Learning outcome 3, a **hazard** is usually defined as anything with the potential to cause harm (chemicals, a fault on electrical equipment, working at height) and a **risk** is the chance (large or small) of harm actually being done when things go wrong (eg risk of electric shock from faulty equipment.

The arrangements for addressing hazards at work start with hazard identification by:

- regularly reviewing the work process or inspecting the site to identify possible hazards
- making an assessment of the possible hazards
- using previous accident and incident reports
- referring to employee reporting, environmental monitoring or testing
- reviewing information from designers, manufacturers, suppliers, consultants or other organisations
- introduction of controls if necessary.

Assessment criteria

4.6 Identify the correct type of fire extinguisher for a particular type of fire.

HOW COMBUSTION TAKES PLACE

Most fires are preventable and, by adopting the right behaviours and procedures, prevention can easily be achieved.

A fire needs three elements to start: a source of ignition (heat), a source of fuel (something that burns) and oxygen. If any one of these elements is removed, a fire will not ignite or will cease to burn.

- Sources of ignition include heaters, lighting, naked flames, electrical equipment, smokers' materials (cigarettes, matches) and anything else that can get very hot or cause sparks.
- Sources of fuel include wood, paper, plastic, rubber or foam, loose packaging materials, waste rubbish, combustible liquids and furniture.
- Sources of oxygen include the air surrounding us.

A fire safety risk assessment using the same approach as used in a health and safety risk assessment should be used to determine the risks.

Based on the findings of the assessment, employers must ensure that adequate and appropriate fire safety measures are in place to minimise the risk of injury or loss of life in the event of a fire. These findings must be kept up to date.

Different classifications of fires

All fires are grouped into classes, according to the type of materials that are burning.

The grid below is a guide to the different types of fire and the type of extinguisher that should be used.

Fire classification	Appropriate type of fire extinguisher			
	Water	Foam	Powder	Carbon dioxide
Class A – Combustible materials such as paper, wood, cardboard and most plastics	✓	✓	✓	
Class B – Flammable or combustible liquids such as petrol, kerosene, paraffin, grease and oil		✓	✓	✓
Class C – Flammable gases, such as propane, butane and methane			✓	✓
Class D – Combustible metals, such as magnesium, titanium, potassium and sodium			Specialist dry powder	
Class F – Cooking oils and fats		Specialist wet chemical		

Fires involving equipment such as electrical circuits or electronic equipment are often referred to as Class E fires, although the category does not officially exist under the BS EN 3 rating system. This is because electrical equipment is often the cause of the fire, rather than the actual type of fire. Most modern fire extinguishers specify on the label whether they should be used on electrical equipment. Normally carbon dioxide or dry powder are suitable agents for putting out a fire involving electricity.

ACTIVITY

Why should you not use water extinguishers on oil fires?

Procedures on discovery of fires on site

How people react in the event of fire depends on how well they have been prepared and trained for a fire emergency. It is therefore imperative that all employees (and visitors and contractors) are familiar with the company procedure to follow in the event of an emergency. A basic procedure is:

- on discovery of a fire, raise the alarm immediately
- if staff are trained and it is considered safe to do so, attempt to fight the fire using the equipment provided
- if fire fighting fails, evacuate (leave the area) immediately
- ensure that no one is left in the fire area and close doors on exit in order to prevent the spread of fire
- go straight to the designated assembly point. These points are specially chosen as they are in locations of safety and where the emergency services are not likely to be obstructed on arrival.

KEY POINT

Remember never to ignore a fire alarm, even if it has gone off many times before that day. It may be the last one you ignore.

Assessment criteria

4.7 Explain situations where asbestos may be encountered.

SmartScreen Unit 301

Handout 32 and Worksheet 32

ASSESSMENT GUIDANCE

Asbestos can be found in some old fuse carriers.

Assessment criteria

4.8 Specify the procedures for dealing with the suspected presence of asbestos in the workplace.

ACTIVITY

In the past, asbestos was often used in fuseboards and as part of the fuse carrier. What action should you take if you discover asbestos during a rewire of an electrical installation?

Abrade

To scrape or wear away.

ASBESTOS ENCOUNTERED IN THE WORKPLACE

Asbestos is the single greatest cause of work-related deaths in the UK. It is a naturally occurring substance, which is obtained from the ground as a rock-like ore, normally through open-pit mining.

Asbestos was used extensively as a building material in the UK from the 1950s through to the mid-1980s. It was used for a variety of purposes and was ideal for fireproofing and insulation. Any building built before the year 2000 (houses, factories, offices, schools, hospitals, etc) may contain asbestos.

Asbestos materials in good condition are safe unless the asbestos fibres become airborne, which happens when materials are damaged due to demolition or remedial works on or in the vicinity of asbestos ceiling tiles, asbestos cement roofs and wall sheets, sprayed asbestos coatings on steel structures and lagging. In older buildings the presence of asbestos in and around boilers, hot water pipes and structural fire protection must always be anticipated when undertaking electrical work. It is difficult to identify asbestos by colour alone and laboratory tests are normally required for positive identification.

WHAT TO DO IF THE PRESENCE OF ASBESTOS IS SUSPECTED

If asbestos is discovered during a work activity, work should be stopped and the employer or duty holder informed immediately.

The Control of Asbestos Regulations 2012 affect anyone who owns, occupies, manages or otherwise has responsibilities for the maintenance and repair of buildings that may contain asbestos. The regulations cover the need for a risk assessment, the need for method statements for the removal and disposal of asbestos, air monitoring and the control measures required. These control measures include personal protective equipment and training.

Implications of being exposed to asbestos

Asbestos commonly comes in the form of chrysotile (white asbestos), amosite (brown asbestos) and crocidolite (blue asbestos). Chrysotile is the common form of asbestos and accounts for 90% to 95% of all asbestos in circulation. When **abraded** or drilled, asbestos produces a fine dust with fibres small enough to be taken into the lungs. Asbestos fibres are very sharp and can lead to mesothelioma (cancer of the lining of the lung), lung cancer, asbestosis (scarring of the lung), diffuse pleural thickening (thickening and hardening of the lung wall) and death.

WHAT YOU NOW KNOW/CAN DO

Learning outcome	Assessment criteria	Page number
1 Understand how relevant health and safety legislation applies in the workplace	*The learner can:*	
	1 specify their own roles and responsibilities and those of others with regard to current relevant legislation	2
	2 specify particular health and safety risks which may be present and the requirements of current health and safety legislation for the range of electrotechnical work operations.	9
2 Understand the procedures for dealing with health and safety in the work environment	*The learner can:*	
	1 state the procedures that should be followed in the case of accidents which involve injury, including requirements for the treatment of electric shock/electrical burns	25
	2 specify appropriate procedures which should be followed when emergency situations occur in the workplace	28
	3 state the limitations of their responsibilities in terms of health and safety in the workplace	30
	4 state the actions to be taken in situations which exceed their level of responsibility for health and safety in the workplace	31
	5 state the procedures that should be followed in accordance with the relevant health and safety regulations for reporting health, safety and/or welfare issues in the workplace	32
	6 specify appropriate responsible persons to whom health and safety and welfare related matters should be reported.	32

Learning outcome	Assessment criteria	Page number
3 Understand the procedures for establishing a safe working environment	*The learner can:*	
	1 state the procedure for producing risk assessments and method statements in accordance with their level of responsibility	34
	2 describe the procedures for working in accordance with provided, pre-determined risk assessments, method statements and safe systems of work.	35
	3 describe the procedures that should be taken to remove or minimise risks before deciding PPE is needed	37
	4 state the purpose of PPE	38
	5 specify the appropriate protective clothing and equipment that is required for identified work tasks	38
	6 state the first aid facilities that must be available in the work area in accordance with health and safety regulations	40
	7 explain why it is important not to misuse first aid equipment/supplies and to replace first aid supplies once used	41
	8 describe safe practices and procedures in the working environment.	41

Learning outcome	Assessment criteria	Page number
4 Understand the requirements for identifying and dealing with hazards in the work environment	*The learner can:*	
	1 identify warning signs for the seven main groups of hazardous substance, as defined by The Chemical (Hazard Information and Packaging for Supply) Regulations (CHIP)	51
	2 define what is meant by the term hazard in relation to health and safety legislation in the workplace	55
	3 identify specific hazards associated with the installation and maintenance of electrotechnical systems and equipment	56
	4 describe situations which can constitute a hazard in the workplace	58
	5 explain practices and procedures for addressing hazards in the work place	68
	6 identify the correct type of fire extinguisher for a particular type of fire	68
	7 explain situations where asbestos may be encountered	70
	8 specify the procedures for dealing with the suspected presence of asbestos in the workplace.	70

ASSESSMENT GUIDANCE

The assessment for this unit is in two sections: a project-based assignment and an e-volve online test.

Assessment A

- The project-based assignment is open book and based around a building with drawings.

Assessment B

- The e-volve paper is a closed book online multiple-choice assessment.
- The paper consists of 20 questions and 40 minutes are allowed.
- You should allow 2 minutes per question.
- Attempt all questions.
- Do not leave until you are confident that you have completed all questions.
- Keep an eye on the time as it seems to move more quickly when you are concentrating.
- Make sure you read each question fully before answering.
- Ensure you know how the e-volve system works. Ask for a demonstration if you are not sure.
- Do not take any paperwork with you into the assessment.
- If you need paper to work anything out, ask the invigilator to provide some.
- Make sure your mobile phone is switched *off* (not on silent) during the assessment. You may be asked to give it to the invigilator.

For Assessment A and Assessment B

- You are allowed to use a scientific calculator (not programmable).
- Keep your work neat and tidy. Don't lose marks due to poor workmanship making it difficult to mark.
- Make sure you arrive in plenty of time for the assessment. Use all the time available – it is not a race to see who finishes first.
- Plan your progress through the exam. Do not spend so much time on one question that you leave insufficient time to finish.
- Do not attempt to copy other candidates' work.

- If you need clarification ask the invigilator. They will not supply technical answers to the questions.
- Go back over your work and check your answers.

Before the assessment

- You will find some questions in the section below to test your knowledge of the learning outcomes.
- Make sure you go over these questions in your own time.
- Spend time on revision in the run-up to the assessment.

OUTCOME KNOWLEDGE CHECK

1   The letters HSE stand for:
    a)   Home Survey Electrical
    b)   Health Survey Epidemic
    c)   Health Safety Executive
    d)   Health Safety Elected.

2   Which one of the following is a statutory (legal) document?
    a)   BS 7671 IET Wiring Regulations 17th edition 2008.
    b)   IET On-Site Guide.
    c)   Electricity at Work Regulations 1989.
    d)   IET Guidance Notes 3.

3   The WEEE Regulations cover the safe disposal of:
    a)   waste of any kind
    b)   waste electrical and electronic equipment
    c)   waste electronic and essential equipment
    d)   waste emergency evacuation extinguishers.

4   The most suitable tool for tightening a 12 mm nut would be:
    a)   half inch spanner
    b)   adjustable wrench
    c)   mole grips
    d)   12 mm spanner.

5   When terminating conductors, which one of the following is good practice?

   a)   Exposed conductors at the terminal.

   b)   Correctly tightened terminals.

   c)   Over long cables in the switchbox.

   d)   Use of strip connector to extend cables.

6   Which one of the following is not a requirement for an electrical installation?

   a)   Complies with specification.

   b)   Complies with BS 7671 IET Wiring Regulations.

   c)   Uses no more than two wiring systems.

   d)   Is safe to use.

7   Upon discovering an unconscious person in contact with live conductors, the first action to take is to:

   a)   go and call your supervisor

   b)   pull the person from the supply

   c)   isolate and make area safe

   d)   apply resuscitation techniques.

8   The use of a smooth file without a handle could result in:

   a)   an uneven finish

   b)   injury to arms or hands

   c)   uneven file wear

   d)   injury to the eyes.

9   The correct type of supply lead, used outside, to supply portable hand tools would be:

   a)   domestic braided flex

   b)   arctic flex

   c)   flat twin

   d)   cat 6.

10 Planned maintenance is carried out to ensure:

   a)   a minimum of disruption to operations

   b)   a maximising of downtime

   c)   avoidance of overtime payments

   d)   a reduction in VAT.

11 An occurence could be defined as 'any incident that could have resulted in an accident but did not'. This would be:

a) an accident

b) a far miss

c) a near miss

d) a lucky escape.

12 PPE suitable for protecting the eyes and ears would be:

a) hard hat and gloves

b) respirator and muffs

c) spats and goggles

d) goggles and muffs.

13 Which of the following PPE options will give the maximum protection to the body when chasing a wall?

a) Apron.

b) Bib and brace overall.

c) Boiler suit.

d) High visibility vest.

14 Which of the following should not be kept in a first aid box?

a) Tablets.

b) Plasters.

c) Bandages.

d) Tweezers.

15 Look at the above sign. What type of sign is this an example of?

a) Warning.

b) Mandatory.

c) Advisory.

d) Prohibition.

16 A water fire extinguisher should only be used on fires such as:

   a) oil

   b) paper

   c) electric

   d) metal.

17 Upon hearing the fire alarm on site, the first action to take is:

   a) continue working until told to leave

   b) find your supervisor for advice

   c) go to the assembly point

   d) leave the site for a tea break.

18 The overriding document covering health and safety at work is the:

   a) Health and Safety at Work Regulations 1974

   b) Health and Safety at Work etc Act 1974

   c) IET Wiring Regulations BS 7671

   d) Manual handling Regulations 2012.

19 The Electricity at Work Regulations allow live working only under certain conditions. One of these would be:

   a) after hours working

   b) accompanied working

   c) electrical testing

   d) quick repairs.

20 To ensure that a section of installation is properly isolated before work is carried out, which of the following should be carried out?

   a) Insulation resistance.

   b) Isolation procedure.

   c) Earth loop test.

   d) Polarity check.

The environment is under ever increasing pressure. The demand for limited natural resources increases year on year. Space for disposal of waste products is becoming more and more difficult to find and much of the waste that is sent to landfill contains materials that are harmful to users of the environment and which can pollute land, air and water.

This unit looks at the legislation that is designed to reduce the impact that work activities have on the environment. It also discusses work methods that help to alleviate pressure on the environment, as well as environmental technologies that are relevant to the construction industry.

The UK government is committed to carbon dioxide (CO_2) emissions reduction in order to address the issue of climate change. The house building sector has been identified as one that can make a significant contribution to this goal of carbon dioxide emissions reduction. It is estimated that 43% of CO_2 emissions in the UK are attributable to domestic properties. The government has introduced the Code for Sustainable Homes, which is the national standard for the sustainable design and construction of new homes. It aims to reduce CO_2 emissions and promote standards of sustainable design higher than the current minimum standards set out by the building regulations. The target is that all new homes will be carbon neutral by 2016. Because of this, any operative working within the construction industry is likely to come into contact with one or more forms of environmental technology systems. A knowledge of the working principles, the advantages and disadvantages, the requirements of the location and an overview of the planning and regulatory requirements for each environmental technology is therefore a prerequisite to working in the construction industry.

LEARNING OUTCOMES

There are three learning outcomes to this unit. The learner will understand:

1 the environmental legislation, working practices and principles which are relevant to work activities

2 how work methods and procedures can reduce material wastage and impact on the environment

3 how and where environmental technology systems can be applied.

The unit will be assessed by:

- online multiple-choice assessment
- one written assignment focusing on PV design and waste management.

Understand the environmental legislation, working practices and principles which are relevant to work activities

Environment

The environment is the land, water and air around us.

WASTE AND THE ENVIRONMENT

In the past, it was common for all waste produced on a construction site to be placed in a skip and for that waste to go to landfill. This practice has resulted in land pollution, ground water pollution and even climate change due to the greenhouse gases that are emitted from landfill sites.

European and UK laws have placed legal obligations on employers and operatives within all industry sectors to reduce waste, avoid pollution, reduce carbon emissions and recycle wherever possible.

What is waste?

Waste is quite difficult to define but, in general terms, it is any item that is thrown away because it is no longer useful or required by its owner. Electrical installation work generates many forms of waste, from packaging materials that come with new equipment and excess materials that cannot be saved for future use, to stripped-out materials and equipment and, of course, general building waste such as brick rubble and timber.

However we define waste, its disposal is governed by legislation. Previously, the majority of construction waste went to landfill sites without any thought to the potential impact of the buried materials on the environment.

European Union laws, that have been applied in the UK, have led to radical changes in waste handling and disposal. If you work within the construction industry, you need to have an understanding of those laws.

What do we mean by the 'environment'?

The Department for Environment Food and Rural Affairs (DEFRA) defines the **environment** as the land, water and air around us. Any pollution of land, water or the air will affect the quality of life for all organisms living within that environment.

Waste and environmental legislation

The environment is under increasing pressure, not only because of our demand for resources, but also due to our need to dispose of waste. Both of these can lead to pollution. There are several legislative documents that determine how we deal with waste and limit our impact on the environment:

- Control of Pollution Act 1974 (COPA)
- Environmental Protection Act 1990 (EPA)
- Environment Act 1995
- The Hazardous Waste Regulations 1995
- Pollution Prevention and Control Act 1999
- The Waste Electrical and Electronic Equipment Regulations 2006 (WEEE Regulations)
- Packaging (Essential Requirements) Regulations 2003

Control of Pollution Act 1974 (COPA)

The aim of this Act is to deal with environmental issues including waste on land, water pollution, air pollution and noise pollution. If a person or organisation is found guilty under this Act, the maximum fine is £5000 plus £50 per day for each day the offence continues after conviction.

Local authorities require construction companies to apply for a permit under the Act prior to starting work. The construction company must carry out an analysis of the likely impact of noise and vibration on those in the surrounding area. The Act gives local authorities the power to impose restrictions on companies or individuals carrying out construction or demolition work, including imposing limits on noise levels and working times so as to avoid causing a nuisance to neighbours.

Environmental Protection Act 1990 (EPA)

The Environmental Protection Act (EPA) applies to England, Scotland and Wales. It deals with the disposal of controlled waste on land and sets out a framework for duty of care. The EPA specifically deals with:

- waste
- contaminated land
- statutory nuisance.

Controlled waste is domestic, commercial and industrial waste – in fact, all waste that is disposed of on the land. Under the EPA, it is an offence for anyone to deposit waste on any land unless a waste management licence authorising that deposit is held.

Land can be contaminated with naturally occurring substances, such as arsenic, or by industrial processes or by fly tipping.

SmartScreen Unit 302
Handout 2

Part 2A of the EPA works on the principle of the 'polluter pays'. The 'polluter' is defined as the person who caused the pollution or who 'knowingly permitted' the contamination. 'Knowingly permitted' does not only apply to allowing the contamination to take place but also to having knowledge of the contamination and failing to deal with it. Where the polluter is unknown then the occupier or owner of the land is responsible.

Part 2A of the EPA applies where significant harm to the land has taken place or where the possibility of significant harm could take place or where rivers or groundwater are or could be affected.

The EPA also covers statutory nuisance and applies to any premises that may be detrimental to health or that cause a nuisance. This section is used by local authorities when dealing with anti-social behaviour, but it also applies to work procedures and covers such things as the emission of:

- dust
- steam
- smells
- **effluvia**
- noise.

When someone complains about any of the above, the local council must investigate. If the investigation reveals that a statutory nuisance does exist, a *notice of abatement* will be issued containing a list of steps that must be followed to reduce the nuisance. In the case of construction, this action could have a serious impact on the completion of the work.

Environment Act 1995

The Environment Act 1995 set new standards for environmental management and led to the creation of a number of agencies to oversee this management. The agencies created by this Act are:

- The Environment Agency
- The Scottish Environment Protection Agency
- The National Park Authorities.

The Act required that the Government prepare strategies on air pollution, national waste and the protection of hedgerows.

The stated purpose of the Environment Agency is to 'enhance or protect the environment and promote sustainable development' and 'to create a better place for people and wildlife'. The Agency looks after everything from fishing rod licences to waste disposal, from flood defences to air pollution.

ASSESSMENT GUIDANCE

It is not only man-made pollution that causes problems. Effluent from farm yards (slurry) causes pollution of waterways, often with disastrous effects on aquatic life.

Effluvia

Emissions of gas, or odorous fumes given off by decaying waste.

The Environment Agency has been given the powers to:

■ stop offending taking place

■ restore and/or remediate, for which it will seek to recover the costs

■ bring under regulatory control

■ punish and/or deter, whether that be by criminal or civil sanctions.

The Environment Agency publishes all prosecutions and associated fines on their website and these range from a couple of thousand pounds for fishing without a licence to many hundreds of thousands of pounds for operating without a waste licence.

The Hazardous Waste Regulations 1995

The Hazardous Waste Regulations set out a regime of control for the tracking and movement of hazardous waste, and deal with the production and disposal of that waste.

Hazardous waste includes such items as:

■ fluorescent tubes

■ television sets

■ fridges

■ PC monitors

■ batteries

■ aerosols and paint

■ contaminated soils.

When hazardous waste is moved from one location to another, a consignment note must be completed and passed to the licensed waste carrier. Hazardous waste must be kept separate from general waste.

Electrical wholesalers generally run schemes whereby fluorescent tubes can be returned to them for safe disposal. It is a requirement of the Hazardous Waste Regulations that records are kept for a period of three years.

Pollution Prevention and Control Act 1999

According to this Act, industries that emit certain substances can only operate with a permit issued by the local authority or the Environment Agency. Included in the schedule of industries requiring a permit are those involved in metal and waste processing.

SmartScreen Unit 302

Handout 2

SmartScreen Unit 302
Handout 4 and Worksheet 4

The Waste Electrical and Electronic Equipment Regulations 2006 (WEEE Regulations)

The WEEE Regulations are the implementation of a European Directive to address the environmental impact of unwanted electrical and electronic equipment, namely to reduce the amount of WEEE that is sent to landfill sites and to encourage recycling, reuse and recovery before disposal in an environmentally friendly manner.

You must comply with the WEEE Regulations if you manufacture, import, rebrand, distribute or dispose of electrical and electronic equipment. While it may seem obvious that manufacturers and distributors must comply, WEEE Regulations apply to *anyone* who disposes of such equipment.

As electricians frequently remove redundant electrical equipment and have surplus materials for disposal, compliance with the WEEE Regulations is an obligation that must be met.

Under the WEEE Regulations, electrical and electronic items are divided into 10 categories:

1   Large household appliances, for example refrigerators, fans and panel heaters
2   Small household appliances, such as vacuum cleaners and toasters
3   IT and telecommunications equipment
4   Consumer equipment, such as radios and televisions
5   Lighting equipment, for example fluorescent tubes and discharge lamps
6   Electrical and electronic tools, such as drills
7   Toys, leisure and sports equipment
8   Medical devices
9   Monitoring and control instruments, such as smoke detectors and thermostats
10 Automatic dispensers, for example vending machines.

Electricians most commonly deal with items in categories 5 and 9 but, at times, other categories may also apply.

Bear in mind that some WEEE may also be classified as hazardous waste. Examples are smoke detectors which contain radioactive emitters and fluorescent tubes which contain mercury and cadmium as well as old discharge lighting control gear, which can contain PCBs (polychlorinated biphenyls), which are also hazardous to persons and the environment. If in doubt regarding any of these items, always seek advice.

ACTIVITY

Look around your own house and name at least five items which would come under the WEEE regulations.

ASSESSMENT GUIDANCE

In years gone by it was common practice for virtually all rubbish to be placed in the dust bin. This is no longer allowed and many items must be sorted into specific categories for recycling.

KEY POINT

The WEEE Regulations apply to electricians when disposing of old electrical equipment.

Packaging (Essential Requirements) Regulations 2003

This Act requires anyone, but generally manufacturers of products, who place packaging into the marketplace, to take certain steps to:

- minimise the amount of packaging used

- make sure packaging can be recovered (recycled)

- ensure that packaging has the minimum possible impact on the environment

- ensure that packaging does not contain high levels of hazardous materials or heavy metals.

Packaging has a very short life cycle – being useful only from the time it leaves the manufacturer to when it arrives at the end user. While the Packaging (Essential Requirements) Regulations are not directly aimed at end users, they do reinforce the requirements for dealing with any waste product.

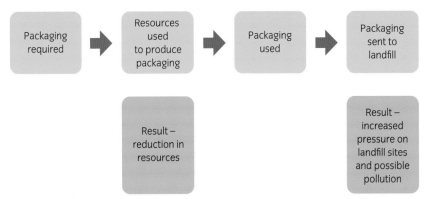

Poor waste model

It is obvious that the poor waste model for packaging is unsustainable. The sustainable waste model shown in the diagram below is a far better model.

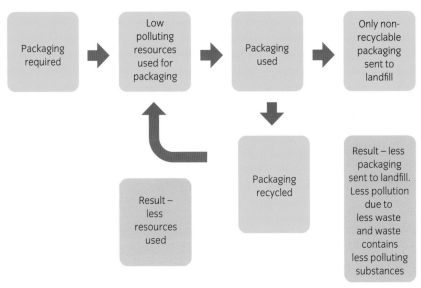

Sustainable waste model

In the sustainable waste model, the packaging is made of materials that will cause minimum pollution in landfill sites. However, the materials are actually recyclable, thus cutting down waste and also reducing the demands on dwindling resources.

The key person in this cycle is the person who ensures that the waste product is recycled.

Assessment criteria

1.3 Describe the ways in which the environment may be affected by work activities

THE IMPACT OF WORK ACTIVITIES

It is true to say that any work activity, especially construction work, will have some effect on the environment. Work activities can affect the land, air and waterways and, in turn, can impact on people and wildlife.

Land

ACTIVITY

State a typical use for turpentine.

The land around a construction site can easily become contaminated with chemicals due to spillages. Common chemicals that are found on construction sites include diesel, petrol, paint, turpentine, white spirit and various oil-based products and preservatives. Any of these, if spilled or poured onto the ground, will result in land contamination. Sending any of the above chemicals or their containers to landfill sites will also result in land contamination.

Other building products, such as asbestos cement sheets, gypsum-based products (such as plasterboard), artex and polystyrene, can be disturbed during building works and should not be sent to general landfill sites.

ASSESSMENT GUIDANCE

Air pollution can be carried a very long way by the prevailing wind, which tends to be from the west in the United Kingdom.

Air pollution

One of the major sources of air pollution is the combustion process. Plant machinery with internal combustion engines will have an effect on the air in and around a construction site. The burning of waste is banned on most sites due to the pollution it creates.

Apart from smoke particles, the burning process may also release harmful gases and chemicals that constitute a health risk or an environmental hazard. As the release of smoke is covered by the EPA, local authorities usually have policies in place that cover bonfires.

Pollution of water courses

Spillage of chemicals poses a great risk to water courses; however, the dumping of contaminated materials in landfill sites can also pollute the water supply. Both wildlife and humans are dependent on clean uncontaminated water for survival; fish stocks are particularly sensitive to water pollution.

BUILDING REGULATIONS AND THE CODE FOR SUSTAINABLE HOMES

While the Building Regulations do not deal with waste products, they are concerned – along with the Code for Sustainable Homes and the EPC (Energy Performance Certificate) for Construction – with reducing carbon emissions, thus protecting the environment. The Building Regulations outline the minimum acceptable standards.

The Code for Sustainable Homes is the national standard for the sustainable design and construction of new homes and classifies homes according to how much energy and water they use. It is a mandatory requirement that all new homes are rated against this standard. The Code contains information on the amount of waste that can be generated during the construction of the building, as a building's classification depends on its design and construction as well as on the future running characteristics of the property. The Code also includes guidelines on such things as external lighting and energy use.

Details of energy efficient lighting and appliance ratings can be found later in this chapter.

HAZARDOUS AND RECYCLABLE WASTE

What is hazardous waste? Hazardous waste is waste that can cause the greatest environmental damage. These wastes are listed in the Lists of Wastes (England) Regulations 2005 and include everyday items such as fluorescent tubes, TVs and computer monitors.

Heavy metals

Much waste is classified as hazardous due to the fact that it contains heavy metals. Heavy metals are natural components of the Earth's crust; small amounts of some heavy metals (such as iron and copper) are beneficial to human life, but arsenic, cadmium, mercury and lead can be poisonous or toxic. These, therefore, pose the greatest pollution risk.

Heavy metals can be introduced into the air, soil or water following dumping of waste materials. The waste materials may be sent to landfill sites where the heavy metals may be dispersed into the air or can contaminate the land; they may, in turn, leach into the water courses or the sea.

Heavy metals have a tendency to **bioaccumulate** rather than pass through the body. This makes them especially dangerous to humans and wildlife. Pockets of concentration, such as may be found at landfill sites, are of the most concern.

Assessment criteria

1.4 Identify and interpret the requirements for electrical installations as outlined in relevant sections of the Building Regulations and the Code for Sustainable Homes

SmartScreen Unit 302
Handout 8 and Worksheet 8

Assessment criteria

1.5 State materials and products that are classed as:

■ hazardous to the environment
■ recyclable

Bioaccumulate

Bioaccumulation occurs when substances such as pesticides or heavy metals gradually build up in the body of an organism, such as a human or other animal. These substances are not flushed through the body and a damaging amount can collect over time.

Plants and animals may absorb these pollutants – this is how heavy metal contamination enters the food chain. For example, heavy metals in sea water are absorbed by small fish, which are eaten by medium-sized fish. These, in turn, are eaten by large fish. The greatest concentration of heavy metals is, therefore, contained in the largest fish in the sea, such as tuna. These are consumed by humans and, thus, the heavy metals accumulate in the human body and can seriously affect health.

Waste sources of heavy metals

■ *Cadmium* is found in nickel–cadmium (NiCad) batteries. Health risks associated with cadmium include an increased risk of bone fractures and defects, and kidney damage.

■ *Lead* is found in old paint, water pipes and old cables. Health risks associated with lead include effects on the gastrointestinal tract, joints, kidneys and the reproductive system, along with acute nerve damage and lowered IQ.

■ *Mercury* is ranked as one of the most severe of polluting waste products and even small amounts can result in neurological damage or death. Excessive exposure to mercury can cause permanent brain or kidney damage. Mercury is found in fluorescent tubes; especially within the fluorescent powder which, if a tube is broken, will disperse into the air and could be inhaled. If old tubes are buried in landfill sites, the mercury can contaminate soil and water. It is estimated that the mercury contained in one fluorescent tube is enough to raise the mercury content of 30,000 litres of water beyond the recognised safe level for drinking water in the UK. Not surprisingly, fluorescent tubes are classified as hazardous waste in the UK.

Recyclable materials

Wherever possible, recycling must be the first option before sending waste to landfill sites. Many recycling and collection schemes exist and construction site managers will be keen for site waste to be sorted and recycled rather than being placed in general skips.

The following materials should be recycled on site:

■ all metallic waste and products containing metal – this includes cable, metallic conduit and trunking, cable support systems and pipework (aluminium, steel and copper are commonly found in the metallic waste generated during the completion of an electrical installation)

■ paper and cardboard – although manufacturers have, in recent years, worked hard to reduce the amount of packaging that comes with products, invariably a large amount of packaging material ends up on site

■ lamps – including fluorescent tubes, sodium, mercury and metal halide discharge lamps.

ACTIVITY

In electrical work, where would the following scrap materials be found:

■ lead
■ steel
■ brass
■ copper
■ aluminium
■ PVC
■ rubber
■ cardboard?

Waste fluorescent tubes

Various recycling schemes, though not specific to electrical installation work, exist for:

- wood
- glass
- plastics, including pipe and conduit
- hardcore (bricks and concrete)
- batteries.

DEALING WITH WASTE

From the discussions in this unit, it is obvious that electrical contractors must have procedures in place to handle, store, recycle and dispose of all waste including hazardous, WEEE and recyclable waste.

Electrical contractors should:

- separate recyclable, hazardous, WEEE and general waste into separate and clearly identified containers
- arrange for collection of recyclable waste by a licensed waste carrier
- arrange for collection and processing of hazardous waste by a licensed waste carrier
- return WEEE to wholesalers or manufacturers or arrange for collection by a licensed waste carrier
- obtain waste transfer notes for all waste
- keep records.

Other procedures are needed to control statutory nuisance and other sources of pollution such as spillage.

ASSESSMENT GUIDANCE

Even small items, such as batteries, should be disposed of properly. Many shops have a collection point for old batteries. Do not put them in the dust bin.

Assessment criteria

1.6 Describe the organisational procedures for processing materials that are classed as:
- hazardous to the environment
- recyclable

ACTIVITY

Produce a simple flow chart to illustrate your company's procedure for dealing with redundant electrical equipment.

OUTCOME 2

Understand how work methods and procedures can reduce material wastage and impact on the environment

Assessment criteria

2.1 State installation methods that can help to reduce material wastage

SmartScreen Unit 302
Handout 14

REDUCING WASTAGE

Whatever the work activity, it will have an impact on the environment. When an item is produced, resources are used up. Extraction of metals and other resources that are vital for the electrical industry impacts directly on the land, air and water, and on humans and wildlife.

Transportation of these resources, energy used in manufacture and transportation of products all involve the production of carbon dioxide. While it is impossible to *eliminate* the impact on the environment, it is possible to *reduce* the impact by reducing the amount of wastage of finished product that occurs.

Methods of reducing waste

Prefabricated materials

Using prefabricated materials reduces the amount of waste material that is generated on site. A number of manufacturers produce custom-designed and prefabricated wiring systems which are basically 'plug-and-play' systems. These not only cut down on waste but also save on labour. They are designed to meet all of the requirements of BS 7671, are fully tested and, thus, cut down on faults.

Careful measurement

The careful measurement of cable length, for example, is one way of reducing wastage. While it is a false economy to cut cables too short, careful measurement will allow cable to be saved rather than leaving long lengths to be cut off and discarded.

The same principle would apply when bending conduit. Careful measurement and skilful practice will enable the operative to get the bend in the right place, minimising the amount of material to be cut off and eliminating the need to recycle mistakes.

Correct sizing

Installing the correct size of circuit cable can help to reduce the unnecessary wastage of precious resources. Following the correct sequence when choosing the right size of cable to install reduces the wastage associated with installation of oversized cables.

ACTIVITY

An electrician installs 45 13 A socket outlets on five ring final circuits with no spurs. If the electrician allows 200 mm per cable at each socket termination and 300 mm for each ring tail at the CCU, how much cable would be wasted?

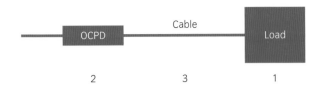

Cable sizing sequence – this figure is traditionally drawn left to right: fuse, cable, load. However, the sequence of sizing is in the order: 1 load, 2 fuse, 3 cable

The steps for cable sizing are:

1 Determine the load, allowing for diversity.

2 Choose the overcurrent protective device (OCPD) so that it is the same rating or larger than the load current.

3 Choose the cable size so that it has the same rating or greater than the OCPD rating.

So the size of cable is determined by the OCPD size which, in turn, is determined by the load. It is quite common for electricians to install circuits for socket outlets as a ring final circuit protected by a 32 A circuit breaker. When asked why, the reply is usually that 'we've always done it that way'. However, it may be that the circuit could have been wired as a radial 16 A circuit due to the size of the assessed load.

Following the correct sequence to determine cable size reduces the burden on the Earth's dwindling resources.

Unnecessary installation of materials

As we have seen, installing the correct size of cable can reduce the amount of cable required. Sometimes a separate earth conductor is installed with a steel wired armour cable whether or not there is a requirement for it. This is of course wasteful if a separate earth conductor is not needed. A sound knowledge of BS 7671 will enable an electrician to know when a separate earth conductor is required and when it is not.

Careful storage

Poor storage can lead to equipment damage and loss of materials. Not only is this wasteful of vital resources but could mean that progress on the job is delayed and that additional cost is involved in sourcing replacements.

To avoid storing materials on site, careful planning is needed. Close cooperation with the supplier or wholesaler is useful as delivery of materials can be arranged for when they are needed. This system, known as 'just in time' (JIT), is used by many large organisations as part of their quality control system.

ASSESSMENT GUIDANCE

Prefabricated wiring systems are not new; they have been around for many years. Some of the early ones were not always that accurate in the measurement of the cables etc.

KEY POINT

Careful design of an installation to reduce voltage drop is another measure to keep energy losses to a minimum.

Reuse

The return of surplus materials to the store will enable those materials to be used on another job, thus reducing both waste and cost and, therefore, increasing profits.

ACTIVITY

What methods of material saving can you apply to your work environment?

Recycle

Whenever possible, waste materials should be recycled rather than being dumped in skips to end up in landfill sites.

SmartScreen Unit 302

Worksheet 14

Recycle

Assessment criteria

2.2 Explain why it is important to report any hazards to the environment that arise from work procedures

REPORTING ENVIRONMENTAL HAZARDS

The Environment Agency has a reporting hotline to enable people to report incidents that have an effect on the environment. These range from the illegal dumping of waste to unlicensed fishing.

On construction sites most risks are controlled. For instance, the burning of waste is banned in most locations. But accidents do happen and the reporting, without delay, of any incident which may cause land, air or water pollution can enable the authorities to prevent a much larger scale environmental problem from developing.

Spillages of oil, diesel, fuel oils, paint, chemicals, battery acids and other liquids, if not dealt with quickly and efficiently, can result in serious land and water pollution which, in turn, could result in damage to wildlife. On construction sites where spillages of such products could take place, a control procedure and action plan should be in place. The first action, on discovering a spillage that could impact on the environment, is to raise the alarm and inform those responsible for the management of the site so that prompt action can be taken to avoid a potential environmental disaster.

CHOOSING ENVIRONMENTALLY FRIENDLY MATERIALS AND PRODUCTS

When planning an electrical installation, consideration not only has to be given to the economic cost of the installation, but to the environmental cost. Along with the reduction of waste and the management of resources, the choice of materials and products also has an impact on the environment.

Energy efficient lighting

Over the past few years there has been a vast improvement in the energy efficiency of lighting. Part L of the Building Regulations makes reference to the Domestic Building Services Compliance Guide: 2010 which requires that internal artificial lighting has a luminous efficacy greater than 45 lamp lumens per circuit watt (lm/W). For a domestic installation, the choice is generally between either compact fluorescent (CF) or LED lighting. LED technology is advancing very quickly and it is expected that **efficacy** of 150 lm/W will be achieved very shortly. **Colour rendition** and the ability to be dimmed have also improved in recent years. There is a further requirement that internal light fittings have a minimum output of 400 lumens.

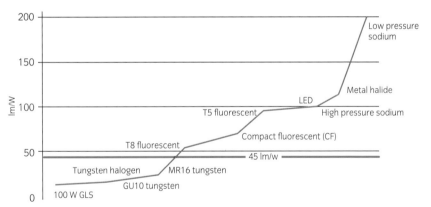

Efficacy of different lamps

For external light fittings the requirements are:

- maximum light fitting output 100 lamp watts

- all lights to be automatically controlled to switch off when area is unoccupied; this can best be achieved by installing passive infrared (PIR) detectors to control the lights

- all lights to be automatically controlled to switch off when daylight is sufficient; this is achieved by means of a photocell (daylight sensor).

Assessment criteria

2.3 Specify environmentally friendly materials, products and procedures that can be used in the installation and maintenance of electrotechnical systems and equipment

Efficacy

The measure of light output of a lamp compared to the energy used by the lamp measured in lumens per watt.

Colour rendition (sometimes called rendering)

The ability of the light emitted by a lamp to show objects in their true colour. For example, some LED lamps have a blue-ish light which can make some objects appear blue.

ASSESSMENT GUIDANCE

Standard tungsten lamps are no longer available due to their low efficacy. When the ban was announced by the government, many people stock piled them as they preferred their brighter light compared to compact fluorescents.

These criteria mean that the maximum rating of tungsten lighting that can be fitted is 100 W. Where greater wattage lighting or manual control is required, the following criteria apply:

■ lamp efficacy must be greater than 45 lumens per circuit watt
■ all lights to be automatically controlled to switch off when daylight is sufficient
■ lighting can be manually controlled by occupants.

The above criteria need to be considered for both new buildings and for extensions to existing buildings where three out of four lights in the areas affected by the building works must meet the above criteria.

Understanding the terminology

To learn how to apply this information regarding lighting, you need to understand the terminology used:

■ *Lumen* (lm) is the SI unit that describes the total amount of visible light emitted by a source.
■ *Luminous efficacy* is a measure of how well a light source produces light and is measured in lumens per watt (lm/W).
■ The *wattage* of a lamp is the rate at which electrical energy is transferred to another form of energy; in the case of lights this transfer is to light and heat.

Example calculation

Does a lamp rated at 42 W and with a light output of 630 lm meet the Building Regulation requirements?

1 Is it rated at less than 100 W?

Yes

2 Is the efficacy 45 lm/W or better? A calculation is needed here:
$$\frac{\text{lumen output}}{\text{wattage}} = \frac{630}{42} = 15 \text{ lm/W}$$

So this lamp does not comply.

Does a lamp rated at 6 W and with a light output of 270 lm meet the Building Regulation requirements?

1 Is it rated at less than 100 W?

Yes

2 Is the efficacy 45 lm/W or better? A calculation is need here:
$$\frac{\text{lumen output}}{\text{wattage}} = \frac{270}{6} = 45 \text{ lm/W}$$

So this lamp complies.

Energy rating system for appliances

Most people will be familiar with the energy rating labels that are displayed on new domestic appliances. This system was introduced in 1992 and is designed to give consumers the ability to make an informed choice when buying appliances. The rating system consists of categories A to G, with A being the most energy efficient. This system has proved to be very successful and manufacturers have striven to improve the efficiency of their products. These days 90% of household appliances achieve an A rating. From June 2011 this system was extended in the UK to include A+, A++ and A+++ ratings to allow for further energy improvements. Choosing appliances with the highest rating possible will result in energy savings.

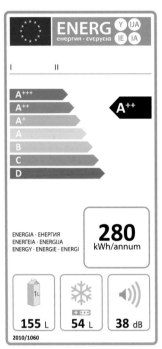

Appliance energy rating label

Building controls

Modern buildings are being designed with environmental controls to reduce energy consumption wherever possible. Occupancy sensors can be fitted to switch off lights, turn down the heating or switch off air conditioning when rooms are unoccupied. Building management systems (BMS) – consisting of environmental sensors which provide information to a central computer that controls lighting, heating and ventilation – are commonly fitted in larger commercial premises. These systems can also control access and fire systems.

Home automation

Many of the control systems installed in commercial premises to save energy are now being seen in domestic premises. Careful control of heating and lighting can lead to big savings in energy bills. Simple sensors such as daylight sensors for outside lights, occupancy sensors for room lights and programmable thermostats fitted to heating systems are effective measures for reducing energy consumption.

Deliver methods to reduce packaging

Rather than having to deal with waste on site, it is far better if waste products do not reach the site in the first place. Packaging is one of the largest sources of on-site waste. Packaging is usually large in size but light in weight, so disposal in a skip does not make environmental or financial sense. Buying bulk-packed materials will cut down on the amount of packaging. Take, for instance, a commercial contract which involves the installation of 30 lay-in luminaires, each having four fluorescent tubes. Each of these fittings may come in a separate box and each tube may come in a separate cardboard sleeve; so there will be 30 boxes and 120 sleeves for disposal or recycling. If the light fittings can be supplied in bulk packaging, the waste will be reduced.

OUTCOME 3

Understand how and where environmental technology systems can be applied

HOW THIS OUTCOME IS ORGANISED

This outcome is divided up according to technology. This has been done to enable the learner to appreciate fully the principles of each technology and how these impact on the installation requirements and the regulatory requirements. The regulatory requirements are covered first, followed by a discussion of the advantages and disadvantages of each technology.

The environmental technologies that are discussed in this outcome are described below.

Heat-producing

- Solar thermal
- Ground source heat pump
- Air source heat pump
- Biomass

Electricity-producing

- Solar photovoltaic
- Micro-wind
- Micro-hydro

Co-generation

- Micro-combined heat and power (heat-led)

Water conservation

- Rainwater harvesting
- Greywater re-use

Environmental technology systems covered

Learning outcome 3 deals with how and where environmental technology systems can be applied and, therefore, the planning requirements and building regulations which apply to each technology. This section explains the terminology used, provides an insight into the workings of both planning and building regulations and explains the differences across the UK.

SmartScreen Unit 302
Handout 19 and Worksheet 19

PLANNING AND PERMITTED DEVELOPMENT

In general, under the Town and Country Planning Act 1990, before any building work that increases the size of a building is carried out, a planning application must be submitted to the local authority. A certain amount of building work is, however, allowed without the need for a planning application. This is known as *permitted development*. Permitted development usual comes with criteria that must be met. When building an extension, for example, it may be possible to do so under permitted development, if the extension is under a certain size, is a certain distance away from the boundary of the property and is not at the front of the property. If the extension does not meet these criteria, then a full application must be made.

The permitted development is intended to ease the burden on local authorities and to smooth the process for the builder or installer. Permitted development exists for renewable technologies and these are outlined within each technology section.

ACTIVITY

Find out who your local authority and water company are. See if you can find any reference to environmental controls on their websites.

BUILDING REGULATIONS

The Climate Change and Sustainable Energy Act 2006 brought micro-generation under the requirements of the building regulations.

Even if a planning application is not required, because the installation meets the criteria for permitted development, there is still a requirement to comply with the relevant building regulations.

Local Authority Building Control (LABC) is the body responsible for checking that building regulations have been met. The person carrying out the work is responsible for ensuring that approval is obtained.

Building regulations are statutory instruments that seek to ensure that the policies and requirements of the relevant legislation are complied with.

The building regulations themselves are rather brief and are currently divided into 14 sections, each of which is accompanied by an approved document. The approved documents are non-statutory and give guidelines on how to comply with the statutory requirements.

The 14 parts of the Building Regulations in England and Wales are listed below.

Part	Title
A	Structure
B	Fire safety
C	Site preparation and resistance to contaminants and moisture
D	Toxic substances
E	Resistance to the passage of sound
F	Ventilation
G	Sanitation, hot-water safety and water efficiency
H	Drainage and waste disposal
J	Combustion appliances and fuel-storage systems
K	Protection from falling, collision and impact
L	Conservation of fuel and power
M	Access to and use of building
N	Glazing – safety in relation to impact, opening and cleaning
P	Electrical safety – dwellings

ASSESSMENT GUIDANCE

Not all green technologies are affected by all sections of the Building Regulations.

There is a 15th Approved Document, which relates to Regulation 7 of the Building Regulations, entitled Approved Document 7 Materials and Workmanship.

Compliance with building regulations is required when installing renewable technologies but not all will be applicable and different technologies will have to comply with different building regulations. Building regulations applicable to each technology are indicated in each of the sections that follow.

ASSESSMENT GUIDANCE

The relationship between the Building Regulations and the approved documents is similar to that between the Electricity at Work Regulations and BS 7671.

DIFFERENCES IN BUILDING REGULATIONS ACROSS THE UK

It should be noted, that due to devolution of government in the UK, each country's government takes responsibility for building regulations, so there are differences between the individual countries.

England and Wales

England and Wales currently follow the same legal structure when it comes to building regulations.

Primary legislation: Building Act 1984

Secondary legislation: Building Regulations 2010

Guidance: 15 Approved Codes of Practice

Although Wales follows the same model as England, the Welsh Government is now responsible for the majority of functions under the Building Act, including making Building Regulations in Wales. The functions that have remained with the UK Government are as set out in The Welsh Ministers (Transfer of Functions) (No. 2) Order 2009 (S.I. 2009/3019).

Scotland

Primary legislation: Building (Scotland) Act 1984

Secondary legislation: Building (Scotland) Regulations 2004 – Amended 2009

Non-statutory guidance: Two Technical Guide Books – Dwellings, Non-dwellings

Northern Ireland

Primary legislation: Building Regulations (Northern Ireland) 1979 Order (Amended 2009)

Secondary legislation: The Building Regulations (Northern Ireland) 2012

Non-statutory guidance: 15 Technical Booklets

The technical booklets have similar content to the approved documents used in England and Wales but the order of the documents is different. A comparison is included overleaf, along with any differences in title.

England & Wales		Northern Ireland	
A	Structure	D	
B	Fire safety	E	
C	Site preparation and resistance to contaminants and moisture	C	
D	Toxic substances	No comparable document	
E	Resistance to the passage of sound	G	
F	Ventilation	K	
G	Sanitation, hot-water safety and water efficiency	P	Sanitary appliances, unvented hot-water storage systems and reducing the risk of scalding
H	Drainage and waste disposal	J	Solid waste in buildings
		N	Drainage
J	Combustion appliances and fuel storage systems	L	
K	Protection from falling, collision and impact	H	Stairs ramps guarding and protection from impact
L	Conservation of fuel and power	F1	Dwellings
		F2	Non-dwellings
M	Access to and use of building	R	
N	Glazing – safety in relation to impact, opening and cleaning	V	
P	Electrical safety – dwellings	No comparable document	
7	Materials and workmanship	B	

Within this unit, reference to parts within the building regulations is applicable to the part designations used in the Building Regulations of England and Wales.

Applying heat-producing micro-renewable energy technologies

SOLAR THERMAL (HOT-WATER) SYSTEMS

A solar thermal hot-water system uses solar radiation to heat water, directly or indirectly.

Working principles

The key components of a solar thermal hot water system are:

1 solar thermal collector

2 differential temperature controller

3 circulating pump

4 hot-water storage cylinder

5 auxiliary heat source.

Assessment criteria

3.1 Describe the fundamental operating principles of solar thermal (hot-water) systems

ASSESSMENT GUIDANCE

A lot of older hot-water systems used gravity to circulate the primary water through the heating coil, because hot water is less dense than cold water. The system shown here uses a pump to circulate the primary water much more quickly.

ACTIVITY

What device is used to move the primary fluid around a solar thermal system?

SmartScreen Unit 302
Handout 23 and Worksheet 23

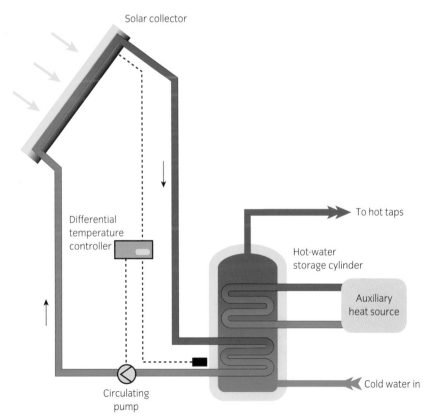

Solar thermal system components

1 Solar thermal collector

A solar thermal collector is designed to collect heat by absorbing heat radiation from the Sun. The heat energy from the Sun heats the heat-transfer fluid contained in the system. Two types of solar collector are used.

Flat-plate collectors are less efficient but cheaper than evacuated tube collectors.

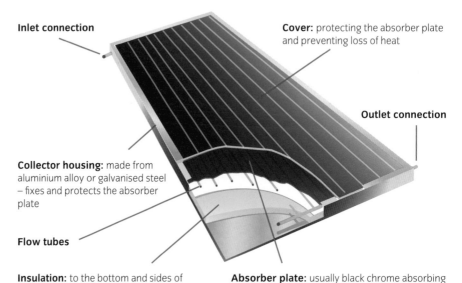

Inlet connection

Cover: protecting the absorber plate and preventing loss of heat

Outlet connection

Collector housing: made from aluminium alloy or galvanised steel – fixes and protects the absorber plate

Flow tubes

Insulation: to the bottom and sides of the collector to reduce the loss of heat

Absorber plate: usually black chrome absorbing coating to maximise heat-collecting efficiency

Cutaway diagram of a flat-plate collector

With this type of collector, the heat-transfer fluid circulates through the collectors and is directly heated by the Sun. The collectors need to be well insulated to avoid heat loss.

Evacuated-tube collectors are more efficient but more expensive than flat-plate collectors.

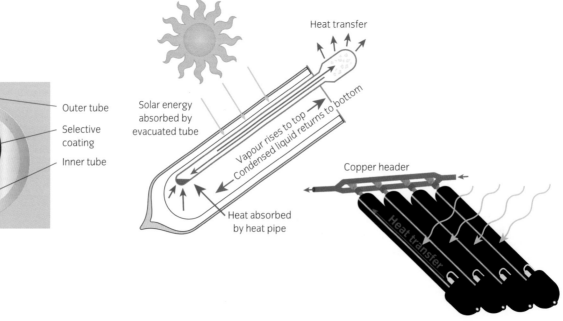

Outer tube

Selective coating

Inner tube

Heat transfer

Solar energy absorbed by evacuated tube

Vapour rises to top
Condensed liquid returns to bottom

Copper header

Heat transfer

Heat absorbed by heat pipe

Evacuated-tube collector

An evacuated-tube collector consists of a specially coated, pressure-resistant, double-walled glass tube. The air is evacuated from the tube to aid the transfer of heat from the Sun to a heat pipe housed within the glass tube. The heat pipe contains a temperature-sensitive medium, such as methanol, that, when heated, vaporises. The warmed gas rises within the tube. A solar collector will contain a number of evacuated tubes linked to a copper header tube that is part of the solar heating circuit. The heat tube is in contact with the header tube. The heat from the methanol vapour in the heat tubes is transferred by conduction to the heat-transfer fluid in the header tube, flowing through the solar heating circuit. This process cools the methanol vapour, which condenses and runs back down to the bottom of the heat tubes, ready for the process to start again. The collector must be mounted at a suitable angle to allow the vapour to rise and the condensed liquid to flow back down the heat pipes.

2 Differential temperature controller

The differential temperature controller (DTC) has sensors connected to the solar collector (high level) and the hot-water storage system (low level). It monitors the temperatures at the two points. The DTC turns the circulating pump on when there is enough solar energy available and there is a demand for water to be heated. Once the stored water reaches the required temperature, the DTC shuts off the circulating pump.

3 Circulating pump

The circulating pump is controlled by the DTC and circulates the system's heat-transfer fluid around the solar hot-water circuit. The circuit is a closed loop between the solar collector and the hot-water storage tank. The heat-transfer fluid is normally water-based but, depending on the system type, usually also contains glycol so that at night, or in periods of low temperatures, it does not freeze in the collectors.

4 Hot-water storage cylinder

The hot-water storage cylinder enables the transfer of heat from the solar collector circuit to the stored water. Several different types of cylinder or cylinder arrangement are possible.

Twin-coil cylinder
With this type of cylinder the lower coil is the solar heating circuit and the upper coil is the auxiliary heating circuit. Cold water enters at the base of the cylinder and is heated by the solar heating coil. If the solar heating circuit cannot meet the required demand, then the boiler will provide heat through the upper coil. Hot water is drawn off, by the taps, from the top of the cylinder.

ASSESSMENT GUIDANCE

When the hot tap is opened the cold-water pressure at the bottom of the tank forces the hot water out at the top.

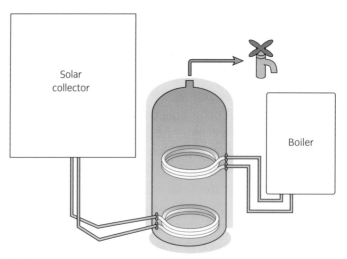

Twin-coil cylinder

Alternatives: One alternative arrangement is to use one cylinder as a solar preheat cylinder, the output of which feeds a hot-water cylinder. The auxiliary heating circuit is connected to the second cylinder.

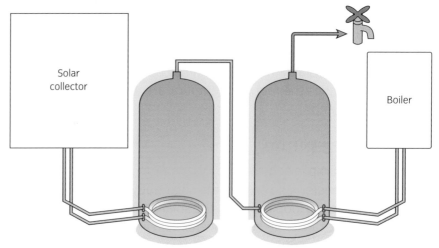

Using two separate cylinders

The two arrangements that have been described are indirect systems, with the solar heating circuit forming a closed loop.

Direct system: An alternative to the indirect system is the direct system, in which the domestic hot water that is stored in the cylinder is directly circulated through the solar collector and is the same water that is drawn off at the taps. Due to this fact, antifreeze (glycol) cannot be used in the system, so it is important to use freeze-tolerant collectors.

5 Auxiliary heat source

In the UK there will be times when there is insufficient solar energy available to provide adequate hot water. On these occasions an auxiliary heat source will be required. Where the premises have space-heating systems installed, the auxiliary heat source is usually this boiler. Where no suitable boiler exists, the auxiliary heat source will be an electric immersion heater.

Location and building requirements

When deciding whether or not a solar thermal hot-water system is suitable for particular premises, the following factors should be considered.

The orientation of the solar collectors

The optimum direction for the solar collectors to face is due south. However, as the Sun rises in the east and sets in the west, any location with a roof facing east, south or west is suitable for mounting a solar thermal system, although the efficiency of the system will be reduced for any system not facing due south.

The tilt of the solar collectors

During the year, the maximum elevation or height of the Sun, relative to the horizon, changes. It is lowest in December and highest in June. Ideally, solar collectors should always be perpendicular to the path of the Sun's rays. As it is generally not practical to change the tilt angle of a solar collector, a compromise angle has to be used. In the UK, the angle is 35°; however, the collectors will work, but less efficiently, from vertical through to horizontal.

Shading of the solar collectors

Any structure, tree, chimney, aerial or other object that stands between the collector and the Sun will block the Sun's energy. The Sun shines for a limited time and any reduction in the amount of heat energy reaching the collector will reduce its ability to provide hot water to meet the demand.

Shading	% of sky blocked by obstacles	Reduction in output
Heavy	> 80%	50%
Significant	60–80%	35%
Modest	20–60%	20%
None or very little	≤ 20%	No reduction

The suitability of the structure for mounting the solar collector

The building structure has to be assessed as to its suitability for the chosen mounting system. Consideration needs to be given to the strength and condition of the structure and the suitability of fixings. The effect of wind must also be taken into account. The force exerted by the wind on the collectors, an upward force known as 'wind uplift', affects both the solar collector fixings and the fixings holding the roof members to the building structure.

ACTIVITY

Which orientation is best for solar thermal systems in the UK?

ACTIVITY

Why is it important to consider the uplift forces on a roof-mounted system, as well as the weight?

In the case of roof-mounted systems on flats and other shared properties, the ownership of the structure on which the proposed system is to be installed must be considered.

The space needed to mount the collectors is dependent on demand for hot water. The number of people occupying premises determines the demand for hot water and, therefore, the number of collectors required and the space need to mount them.

Compatibility with the existing hot-water system

Solar thermal systems provide stored hot water rather than instantaneous hot water.

- Premises using under/over-sink water heaters and electric showers will not be suitable for the installation of a solar thermal hot-water system.

- Premises using a combination boiler to provide hot water will not be suitable for the installation of a solar thermal hot-water system unless substantial changes are made to the system.

Planning permission

Permitted development applies where a solar thermal system is installed:

- on a dwelling house or block of flats
- on a building within the grounds of a dwelling house or block of flats
- as a stand-alone system in the grounds of a dwelling house or block of flats.

However there are criteria to be met in each case.

For building mounted systems:

- the solar thermal system cannot protrude more than 200 mm from the wall or the roof slope
- the solar thermal system cannot protrude past the highest point of the roof (the ridgeline), excluding the chimney.

The criteria that must be met for stand-alone systems are that:

- only one stand-alone system is allowed in the grounds
- the array cannot exceed 4 m in height
- the array cannot be installed within 5 m of the boundary of the grounds
- the array cannot exceed 9 m² in area
- no dimension of the array can exceed 3 m.

For both stand-alone and building mounted systems:

- the system cannot be installed in the grounds or on a building within the grounds of a listed building or a scheduled monument
- if the dwelling is in a conservation area or a World Heritage Site, then the array cannot be closer to a highway than the house or block of flats.

In cases that fall outside permitted development, planning permission will be required.

Compliance with building regulations

The following building regulations will apply to solar thermal hot-water systems.

Part	Title	Relevance
A	Structure	Where solar collectors and other components can put extra load on the structure, particularly the roof structure, not only the additional downwards load but also the uplift caused by the wind must be considered.
B	Fire safety	Where holes for pipes are made, this may reduce the fire resistance of the building fabric.
C	Resistance to contaminants and moisture	Where holes for pipes and fixings for collectors are made, this may reduce the moisture resistance of the building and allow ingress of water.
G	Sanitation, hot-water safety and water efficiency	Hot-water safety and water efficiency
L	Conservation of fuel and power	Energy efficiency of the system and the building as a whole
P	Electrical safety	The installation of electrical controls and components

Other regulatory requirements to consider

- BS 7671 Requirements for Electrical Installations
- Approved document Part G3: Unvented hot-water storage systems
- Water Regulations (WRAS)

Assessment criteria

3.2 State the application and limitations of solar thermal (hot water) systems

Advantages

- It reduces CO_2 emissions.
- It reduces energy costs.
- It is low maintenance.
- It improves the energy rating of the building.

Disadvantages

- It may not be compatible with the existing hot-water system.
- It may not meet demand for hot water in the winter.
- There are high initial installation costs.
- It requires a linked auxiliary heat source.

Assessment criteria

3.1 Describe the fundamental operating principles of heat pumps

HEAT PUMPS

Working principles

A water pump moves water from a lower level to a higher level, through the application of energy. Pumping the handle draws water up from a lower level to a higher level through the application of kinetic energy.

As the name suggests, a heat pump moves heat energy from one location to another by the application of energy. In most cases, the applied energy is electrical energy.

Heat energy from the Sun exists in the air that surrounds us and in the ground beneath our feet. At *absolute zero* or 0 K (kelvin), there is no heat in a system. This temperature is equivalent to −273 °C, so, even with outside temperatures of −10 °C, there is a vast amount of free heat energy available.

−273		−10	0		20

Heat energy exists down to absolute zero (0 K ≈ −273 °C)

A heat pump moves heat from one location to another, just as a water pump moves water from one location to another.

By using a relatively small amount of additional energy, that stored heat energy in the air or in the ground can be extracted and put to use in heating our living accommodation.

Heat pumps extract heat from outside and transfer it inside, in much the same way that a refrigerator extracts heat from the inside of the refrigerator and releases it at the back of the refrigerator via the heat-exchanger fins.

A basic rule of heat transfer is that heat moves from warmer spaces to colder spaces.

ASSESSMENT GUIDANCE

Look at your refrigerator or freezer at home. The inside is cold because the heat energy has been removed from it. The tubes at the back are hot.

A heat pump contains a refrigerant. The external air or ground is the medium or heat source that gives up its heat energy. When the refrigerant is passed through this heat source the refrigerant is cooler than its surroundings and so absorbs heat causing the refrigerant to evaporate or turn to gas. The compressor on the heat pump then compresses the refrigerant gas, causing the gas to heat up. When the refrigerant is passed to the interior, the refrigerant is now hotter than its surroundings and gives up its heat to the cooler surroundings. The refrigerant is then allowed to expand, where it condenses back into a liquid. This cools the refrigerant and the cycle starts again.

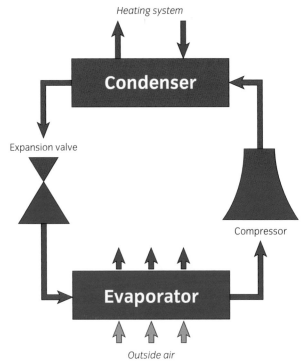

The refrigeration process

The only energy needed to drive the system is what is required by the compressor. The greater the difference in temperature between the refrigerant and the heat-source medium from where heat is being extracted, the greater the efficiency of the heat pump. If the heat-source medium is very cold then the refrigerant will need to be colder, to be able to absorb heat, so the harder the compressor must work and the more energy is needed to accomplish this.

The two main types of heat pump in common use are:

■ ground source heat pumps (GSHP)

■ air source heat pumps (ASHP).

Heat pumps extract heat energy from the air or the ground, but the energy extracted is replaced by the action of the Sun.

It is not uncommon for heat pumps to have efficiencies in the order of 300%; for an electrical input of 3 kW, a heat output of 9 kW is achievable. If we compare this to other heat appliances we can see where the savings are made.

Electricity input 1 kW

Heat output 1 kW

100% efficiency

The efficiency of an electric panel heater

Gas input 1 kW

Heat output 0.95 kW

95% efficiency

The efficiency of an A-rated condensing gas boiler

Electricity input 1 kW

+2 kW free heat extracted from air

Heat output 3 kW

equates to a 300% efficiency

The efficiency of an air source heat pump (image courtesy of Bosch Thermotechnology Limited)

ACTIVITY

Visit the website of a gas boiler supplier and compare the efficiency of a conventional boiler to that of a condensing gas boiler.

The efficiency of a heat pump is measured in terms of the *coefficient of performance* (COP), which is the ratio between the heat delivered and the power input of the compressor.

$$COP = \frac{\text{heat delivered}}{\text{compressor power}}$$

The higher the COP value, the greater the efficiency. Higher COP values are achieved in mild weather than in cold weather because, in cold weather, the compressor has to work harder to extract heat.

Storing excess heat produced

Heat pumps are not able to provide instant heat and so therefore work best when run continuously. Stop–start operations will shorten the lifespan of a heat pump.

A buffer tank, simply a large water-storage vessel, is incorporated into the circuit so that, when heat is not required within the premises, the heat pump can 'dump' heat to it and thus keep running. When there is a need for heat, this can be drawn from the buffer tank. A buffer tank can be used with both ground source and air source heat pumps.

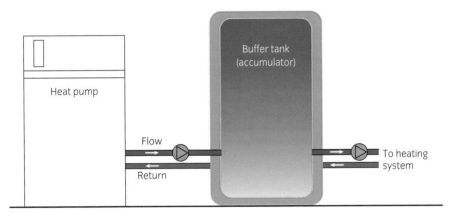

Storing heat in a buffer tank

GROUND SOURCE HEAT PUMPS

Assessment criteria

3.1 Describe the fundamental operating principles of ground source heat pumps

Working principles

A ground source heat pump (GSHP) extracts free heat at low-temperature, from the ground, upgrades it to a higher temperature and releases it, where required, for space heating and water heating.

ASSESSMENT GUIDANCE

Although the ground or air may 'feel' cold, there is still energy available to be collected.

The key components of a GSHP are:

- heat-collection loops and a pump
- heat pump
- heating system.

The collection of heat from the ground is accomplished by means of pipes containing a mixture of water and antifreeze, which are buried in the ground. This type of system is known as a 'closed-loop' system.

Three methods of burying the pipes are used, with each having advantages and disadvantages.

Horizontal loops

Piping is installed in horizontal trenches that are generally 1.5–2 m deep. Horizontal loops require more piping than vertical loops – around 200 m of piping for the average house.

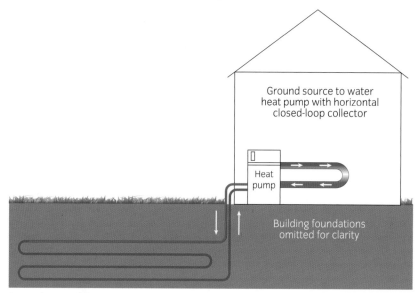

Horizontal ground loops

Vertical loops

Most commercial installations use vertical loops. Holes are bored to a depth between 15–60 m, depending on soil conditions, and spaced approximately 5 m apart. Pipe is then inserted into these bore holes. The advantage of this system is that less land is needed.

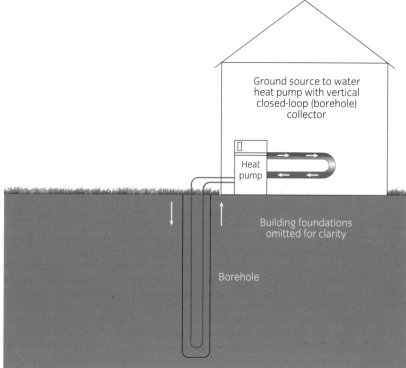

Vertical ground loops

ACTIVITY

What ground structure/material would make vertical loops impracticable or very expensive?

Slinkies

Slinky coils are flattened, overlapping coils that are spread out and buried, either vertically or horizontally. They are able to concentrate the area of heat transfer into a small area of land. This reduces the length of trench needing to be excavated and therefore the amount of land required. Slinkies installed in a 10 m long trench will yield around 1 kW of heating load.

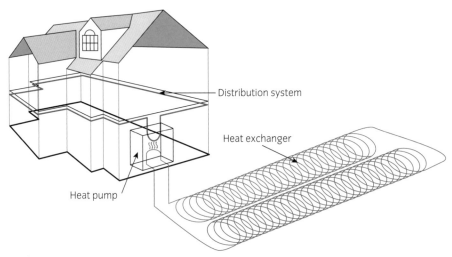

Slinkies

Slinkies being installed in the ground

ASSESSMENT GUIDANCE

Looking at this photograph, it is obvious that unless you want your garden to be given the 'Time Team' treatment it is best to have the ground source heat pump pipes installed during the original building stage.

The water–antifreeze mix is circulated around these ground pipes by means of a pump. The low-grade heat from the ground is passed over a heat exchanger, which transfers the heat from the ground to the refrigerant gas. The refrigerant gas is compressed and passed across a second heat exchanger, where the heat is transferred to a pumped heating loop that feeds either radiators or under-floor heating.

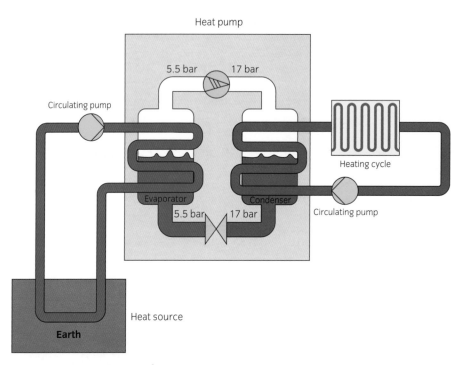

Ground source heat pump operating principle

The water output from the GSHP is at a lower temperature than would be obtained from a gas boiler.

GSHP heat output is at 40 °C, compared to a gas boiler at 60–80 °C.

For this reason, under-floor heating, which requires temperatures of 30–35 °C, is the most suitable form of heating arrangement to use with a GSHP. Low-temperature or oversized radiators could also be used.

A GSHP system in itself is unable to heat hot water directly to a suitable temperature. Hot water needs to be stored at a temperature of 60 °C. An ancillary heating device will be required to reach the required temperatures.

A GSHP is unable to provide instant heat and, for maximum efficiency, should run all the time. In some cases it is beneficial to fit a buffer tank to the output so that any excess heat is stored, ready to be used when required.

By reversing the refrigeration process, a GSHP can also be used to provide cooling in summer.

ACTIVITY

What is another name for a buffer tank?

Location and building requirements

For a GSHP system to work effectively, and due to the fact that the output temperature is low, the building has to be well insulated.

A suitable amount of land has to be available for trenches or, alternatively, land that is suitable for bore holes. In either case, access for machinery will be required.

Planning permission

The installation of a ground source heat pump is usually considered to be permitted development and will not require a planning application to be made.

If the building is a listed building or in a conservation area, the local area planning authority will be need to be consulted.

Compliance with building regulations

The following building regulations will apply.

Part	Title	Relevance
A	Structural safety	Where heat pumps and other components put additional load on the building structure or where openings are formed to pass from outside to inside
B	Fire safety	Where holes for pipes may reduce the fire-resistant integrity of the building structure
C	Resistance to contaminants and moisture	Where holes for pipes may reduce the moisture-resistant integrity of the building structure
E	Resistance to sound	Where holes for pipes may reduce the soundproof integrity of the building structure
G	Sanitation, hot-water safety and water efficiency	Hot-water safety and water efficiency Unvented hot-water system
L	Conservation of fuel and power	Energy efficiency of the system and the building as a whole
P	Electrical safety	The installation of electrical controls and components as well as the supply to the heat pump

ASSESSMENT GUIDANCE

It is not good practice to install a heat pump via a plug and socket. Any load over 3 kW (13 A) will need its own circuit wired back to the consumer unit by a competent electrician.

Other regulatory requirements to consider

- BS 7671 Requirements for Electrical Installations
- F (fluorinated) gas requirements if working on refrigeration pipework

Advantages

- There is high efficiency.
- There is a reduction in energy bills – cheaper to run than electric, gas or oil boilers.
- There is a reduction in CO_2 emissions.
- There are no CO_2 emissions on site.
- They are safe, as no combustion takes place and no emission of potentially dangerous gases.
- They are low maintenance, compared with combustion devices.
- They have a long lifespan.
- There is no requirement for fuel storage, so less installation space is required.
- They can be used to provide cooling in summer.
- They are more efficient than air source heat pumps.

Disadvantages

- The initial costs are high.
- They require large ground area or boreholes.
- The design and installation are complex tasks.
- They are unlikely to work efficiently with an existing heating system.
- They use refrigerants, which could be harmful to the environment.
- They are more expensive to install than air source heat pumps.

ACTIVITY

Why is it so important to reduce our output of CO_2?

AIR SOURCE HEAT PUMPS

Working principles

An air source heat pump (ASHP) extracts free heat from low-temperature air and releases it where required, for space heating and water heating.

The key components of an ASHP are:

- a heat pump containing a heat exchanger, a compressor and an expansion valve
- a heating system.

Air source heat pump

An ASHP works in a similar way to a refrigerator, but the cooled area becomes the outside world and the area where the heat is released is the inside of a building. The steps of the ASHP process are as follows.

- The pipes of the pump system contain refrigerant that can be a liquid or a gas, depending on the stage of the cycle. The refrigerant, as a gas, flows through a heat exchanger (evaporator), where low-temperature air from outside is drawn across the heat exchanger by means of the unit's internal fan. The heat from the air warms the refrigerant. Any liquid refrigerant boils to gas.

- The warmed refrigerant vapour then flows to a compressor, where it is compressed, causing its temperature to rise further.

- Following this pressurisation stage, the refrigerant gas passes through another heat exchanger (condenser), where it loses heat to the heating-system water, because it is hotter than the system water. At this stage, some of the refrigerant has condensed to a liquid. The heating system carries heat away to heat the building.

- The cooled refrigerant passes through an expansion valve, where its pressure drops suddenly and its temperature falls. The refrigerant flows as a liquid to the evaporator heat exchanger, continuing the cycle.

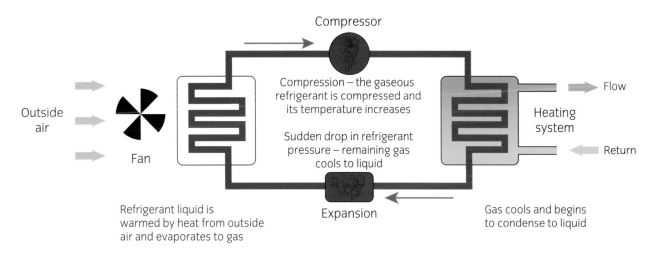

Air source heat pump operating principle

The two types of ASHP in common use are:

- air-to-water – the type described above, which can be used to provide both space heating and water heating
- air-to-air – this type is not suitable for providing water heating.

The output temperature of an ASHP will be lower than that of a gas-fired boiler. Ideally, the ASHP should be used in conjunction with an under-floor heating system. Alternatively, it could be used with low-temperature radiators.

Assessment criteria

3.2 State the applications and limitations of air source heat pump systems

ASSESSMENT GUIDANCE

Older houses in the UK are generally not well insulated. Those with solid walls obviously cannot have cavity wall insulation, but insulation can still be fitted. Loft insulation is the most effective and there are government schemes and grants available for the supply and installation of the insulation material.

Assessment criteria

3.3 State the Local Authority Building Control requirements which apply to the installation of air source heat pumps

ACTIVITY

Identify suitable insulation methods for:

a) roof space
b) cavity walls
c) windows
d) hot water pipes.

Location and building requirements

When deciding whether or not an ASHP is suitable for the premises, the following should be considered.

- The premises must be well insulated.
- There must be space to fit the unit on the ground outside the building, or mounted on a wall. There will also need to be clear space around the unit to allow an adequate airflow.
- The ideal heating system to couple to an ASHP is either under-floor heating or warm-air heating.
- An ASHP will pay for itself in a shorter period of time if it replaces an electric, coal or oil heating system than if it is replacing a gas-fired boiler.

Air sourced heat pumps are an ideal solution for new-build properties, where high levels of insulation and under-floor heating are to be installed.

Planning permission

Permitted development applies where an air source heat pump is installed:

- on a dwelling house or block of flats
- on a building within the grounds of a dwelling house or block of flats
- in the grounds of a dwelling house or block of flats.

There are, however, criteria to be met, mainly due to noise generation by the ASHP.

- The air source heat pump must comply with the MCS Planning Standards or equivalent.
- Only one ASHP may be installed on the building or within the grounds of the building.
- A wind turbine must not be installed on the building or within the grounds of the building.
- The volume of the outdoor unit's compressor must not exceed $0.6\,m^3$.
- It cannot be installed within 1 m of the boundary.
- It cannot be installed on a pitched roof.
- If it is installed on a flat roof, it must not be within 1 m of the roof edge.
- If installed on a wall that fronts a highway, it cannot be mounted above the level of the ground storey.
- It cannot be installed on a site designated as a monument.

- It cannot be installed on a building that is a listed building, or in its grounds.
- It cannot be installed on a roof or a wall that fronts a highway, or within a conservation area or World Heritage Site.
- If the dwelling is in a conservation area or a World Heritage Site, then the ASHP cannot be closer to a highway than the house or block of flats.

Compliance with building regulations

The following building regulations will apply.

Part	Title	Relevance
A	Structural safety	Where heat pumps and other components put additional load on the building structure, for instance, where the heat pump is installed on the roof or on a wall
B	Fire safety	Where holes for pipes may reduce the fire-resistant integrity of the building structure
C	Resistance from contaminants and moisture	Where holes for pipes may reduce the moisture-resistant integrity of the building structure
E	Resistance to sound	Where holes for pipes may reduce the soundproof integrity of the building structure
G	Sanitation, hot-water safety and water efficiency	Hot-water safety and water efficiency
L	Conservation of fuel and power	Energy efficiency of the system and the building as a whole
P	Electrical safety	The installation of electrical controls and components as well as the supply to the heat pump

Other regulatory requirements to consider

- BS 7671 Requirements for Electrical Installations.
- F (fluorinated) Gas Regulations if working on refrigeration pipework.

Advantages

- There is high efficiency.
- There is a reduction in energy bills – they are cheaper to run than electric, gas or oil boilers.
- There is a reduction in CO_2 emissions.
- There are no CO_2 emissions on site.
- They are safe, as no combustion takes place and there is no emission of potentially dangerous gases.
- They are low maintenance, compared with combustion devices.
- There is no requirement for fuel storage, so less installation space is required.
- They can be used to provide cooling in summer.
- They are cheaper and easier to install than GSHPs.

Disadvantages

- They are unlikely to work efficiently with an existing heating system.
- They are not as efficient as GSHPs.
- The initial cost is high.
- They are less efficient in winter.
- There is noise from fans.
- They have to incorporate a defrost cycle to stop the heat exchanger freezing in winter.

BIOMASS

What is biomass? Biomass is biological material from living or recently living organisms. Biomass fuels are usually derived from plant-based material but could be derived from animal material.

The major difference between biomass and fossil fuels, both of which are derived from the same source, is time. Fossil fuels, such as gas, oil and coal, have taken millions of years to form. Demand for these fuels is outstripping supply and replenishment. Biomass is derived from recently living organisms. As long as these organisms are replaced by replanting, and demand does not outstrip replacement time, the whole process is sustainable. Biomass is therefore rightly regarded as a renewable energy technology.

Both fossil fuels and biomass fuels are burnt to produce heat and both produce carbon dioxide. This is a greenhouse gas that has been linked to global warming. During their lives, plants and trees absorb carbon dioxide from the atmosphere, to enable growth to take place. When these plants are burnt, the carbon dioxide is released once again into the atmosphere.

So how does biomass have a carbon advantage over fossil fuels? The answer again is time. Fossil fuels absorbed carbon dioxide from the atmosphere millions of years ago and have trapped that carbon dioxide ever since. When fossil fuels are burnt, they release the carbon dioxide from all those millions of years ago and so add to the present-day atmospheric carbon dioxide level.

Biomass absorbs carbon dioxide when it grows, reducing current atmospheric carbon dioxide levels. When biomass is turned into fuel and burnt, it releases the carbon dioxide back into the atmosphere. The net result is that there is no overall increase in the amount of carbon dioxide in the atmosphere.

ASSESSMENT GUIDANCE

Biomass is in some ways a further development from the wood-burning back boiler, with more controls.

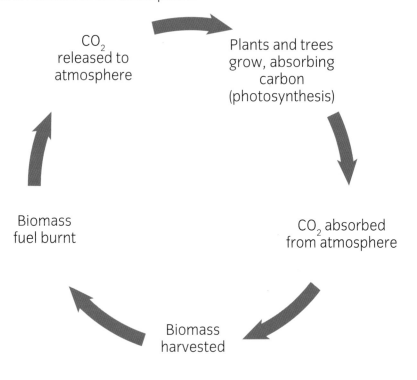

The carbon cycle

A disadvantage of biomass is that the material is less dense than fossil fuels so, to achieve the same heat output, a greater quantity of biomass than fossil fuel is required. However, with careful management, the use of biomass is sustainable, whereas the use of fossil fuels is not.

The classes of biomass raw material that can be turned into biomass fuels are:

- wood
- crops such as elephant grass, reed canary grass and oil-seed rape
- agricultural by-products such as straw, grain husks, forest product waste, animal waste such as chicken litter and slurry
- food waste – it is estimate that some 35% of food purchased ends up as waste
- industrial waste.

Woody biomass

For domestic use, wood-related products are the primary biomass fuels.

For wood to work as a sustainable material, the trees used need to be relatively fast growing, so short-rotation coppice woodlands containing willow, hazel and poplar are used on a 3–5 year rotation. Because of this, large logs are not available, neither can slow-growing timbers that would have a higher calorific value be used. Woody biomass as a fuel is generally supplied as:

- small logs
- wood chips – mechanically shredded trees, branches, etc
- wood pellets – formed from sawdust or shavings that are compressed to form pellets.

The *calorific value* (energy given off by burning) of woody biomass is generally low. The greener (wetter) the wood is, the lower the calorific value will be.

Woody biomass boilers

A biomass boiler can be as simple as a log-burner providing heat to a single room or may be a boiler heating a whole house.

Woody biomass boilers can be automated so that a constant supply of fuel is available. Wood pellets are transferred to a combustion chamber by means of an auger drive or, if the fuel storage is remote from the boiler, by a suction system. The combustion process is monitored via thermostats in the flue gases and adjustments are made to the fan speed, which controls air intake, and to the fuel-feed system, to control the feed of pellets. All of this is controlled by a microprocessor.

The hot flue gases are passed across a heat exchanger, where the heat is transferred to the water in the central-heating system. From this point the heated water is circulated around a standard central-heating system.

In automated, biomass boilers, heat exchangers are self-cleaning and the amount of ash produced is relatively small. As a result, the boilers require little maintenance. The waste gases are taken away from the boiler by the flue and are then dispersed via the flue terminal.

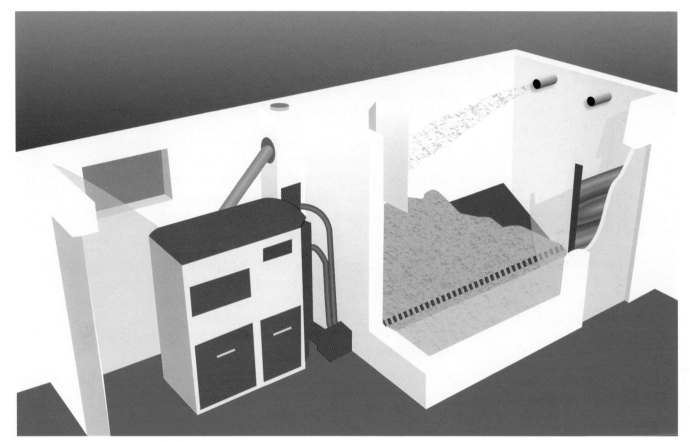

Biomass boiler with suction feed system

Location and building requirements

When considering the installation of a biomass boiler, the following considerations should be taken into account.

- Space will be required for storage of biomass fuel.
- Easy access will be required for delivery of biomass fuel.
- A biomass boiler may not be permitted in a designated smokeless zone.

Smoke-control areas and exempt appliances

In the past, when it was common to burn coal as a source of domestic heat or for the commercial generation of heat and power, many cities suffered from very poor air quality and *smogs*, which are a mixture of winter fog and smoke, were common.

These smogs contained high levels of sulphur dioxide and smoke particles, both of which are harmful to humans.

In December 1952, a period of windless conditions prevailed, resulting in a smog in London that lasted for five days. Apart from very poor visibility, it was estimated at the time that this smog resulted in some 4000 premature deaths and another 100 000 people suffered smog-related illnesses.

Recent reports have found that these figures were seriously underestimated and as many as 12 000 may have died.

Not surprisingly, there was a massive public outcry, which lead to the government introducing the Clean Air Act 1956 and local authorities declaring areas as 'smoke-control areas'.

The Clean Air Act of 1956 was replaced in 1993. Under the Clean Air Act 1993 it is an offence to sell or burn an *unauthorised fuel* in a smoke control area unless it is burned in what is known as an 'exempt appliance'. These appliances are able to burn fuels that would normally be 'smoky', without emitting smoke to the atmosphere. Each appliance is designed to burn a specific fuel.

Lists of authorised fuels and exempt appliances can be found on the Department for the Environment Food and Rural Affairs (Defra) website.

Planning permission

Planning permission will not normally be required for the installation of a biomass boiler in a domestic dwelling if all of the work is internal to the building.

If the installation requires an external flue to be installed, it will normally be classed as permitted development as long as the following criterion is met.

- The flue is to the rear or side elevation and does not extend more than 1 m above the highest part of the roof.

Listed building or buildings in a designated area

Check with the local planning authority for both internal work and external flues.

Buildings in a conservation area or in a World Heritage site

Flues should not be fitted on the principle or side elevation if they would be visible from a highway.

If the project includes the construction of buildings for storage of the biofuels, or to house the boiler, then the same planning requirements as for extensions and garden outbuildings will apply.

KEY POINT

Before fitting a biomass boiler in a smoke control area it is necessary to confirm that it is classified as an exempt appliance by Defra.

Assessment criteria

3.3 State the Local Authority Building Control requirements which apply to the installation of biomass technology

KEY POINT

Before deciding on any particular system, contact the local authority to see if there are any special requirements.

Compliance with building regulations

The following building regulations will apply.

Part	Title	Relevance
A	Structural safety	Where the biomass appliance and other components put load on the structure
B	Fire safety	Where holes for pipes, etc. may reduce the fire-resisting integrity of the building structure
C	Resistance to contaminants and moisture	Where holes for pipes, etc may reduce the moisture-resisting integrity of the building structure
E	Resistance to sound	Where holes for pipes, etc may reduce the soundproof integrity of the building structure
G	Sanitation, hot-water safety, and water efficiency	Hot-water safety and water efficiency
J	Heat-producing appliances	Biomass boilers produce heat and therefore must be installed correctly
L	Conservation of fuel and power	Energy efficiency of the installed system and the building
P	Electrical safety	Safe installation of the electrical supplies and any controls

Advantages

- It is carbon neutral.
- It is a sustainable fuel source.
- When it is burnt, the waste gases are low in nitrous oxide, with no sulphur dioxide – both are greenhouse gases.

Assessment criteria

3.2 State the applications and limitations of biomass systems

ACTIVITY

What is meant by the term 'greenhouse gas'?

Disadvantages

- Transportation costs are high – wood pellets or chips will need to be delivered in bulk to make delivery costs viable.

- Storage space is needed for the fuel. As woody biomass has a low calorific value, a large quantity of fuel will be required. Consideration must be given to whether or not adequate storage space is available.

- Control – when a solid fuel is burnt, it is not possible to have instant control of heat, as would be the case with a gas boiler. The fuel source cannot be instantly removed to stop combustion.

- It requires a suitable flue system.

ASSESSMENT GUIDANCE

Unless a fully automated system is used, there can be quite a lot of manual handling associated with biomass systems.

Applying electricity-producing micro-renewable energy technologies

The electricity-producing micro-renewable energy technologies that will be discussed in this section are:

- solar photovoltaic
- micro-wind
- micro-hydro.

The major advantages of these technologies are that they do not use any of the planet's dwindling fossil fuel resources. They also do not produce any carbon dioxide (CO_2) when running.

With each of the electricity producing micro-renewable energy technologies, two types of connection exist:

1 on-grid or grid tied – where the system is connected in parallel with the grid supplied electricity

2 off-grid – where the system is not connected to the grid but supplies electricity directly to current-using equipment or is used to charge batteries and then supplies electrical equipment via an inverter.

The batteries required for off-grid systems need to be deep-cycle type batteries, which are expensive to purchase. The other downside of using batteries to store electricity is that the batteries' life span may be as short as five years, after which the battery bank will require replacing.

With on-grid systems, any excess electricity generated is exported back to the grid. At times when the generation output is not sufficient to meet the demand, electricity is imported from the grid.

While the following sections will be focused primarily on on-grid or grid-tied systems, which are the most common type in use, an overview of the components required for off-grid systems is included to provide a complete explanation of the technology.

SmartScreen Unit 302
Handout 27 and Worksheet 27

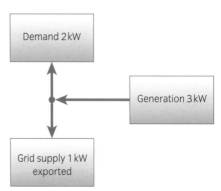

Generation exceeds demand

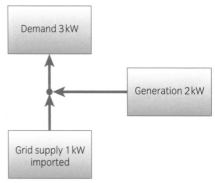

Demand exceeds generation

Assessment criteria

3.1 Describe the fundamental operating principles of solar photovoltaic technology

SOLAR PHOTOVOLTAIC (PV)

Solar photovoltaic (PV) is the conversion of light into electricity. Light is electromagnetic energy and, in the case of visible light, is electromagnetic energy that is visible to the human eye. The electromagnetic energy released by the Sun consists of a wide spectrum, most of which is not visible to the human eye and cannot be converted into electricity by PV modules.

ASSESSMENT GUIDANCE

Photovoltaic is probably the most common system being installed. It is fairly simple to install and requires minimum disruption when connecting the system to the meter position.

Working principles

The basic element of photovoltaic energy production is the PV cell, which is made from semiconductor material. A semiconductor is a material with resistivity that sits between that of an insulator and a conductor. Whilst various semiconductor materials can be used in the making of PV cells, the most common material is silicon. Adding a small quantity of a different element (an impurity) to the silicon, a process known as 'doping', produces n-type or p-type semiconductor material. Whether it is n-type (negative) or p-type (positive) semiconductor material is dependent on the element used to dope the silicon. Placing an n-type and a p-type semiconductor material together creates a p-n junction. This forms the basis of all semiconductors used in electronics.

When *photons*, which are particles of energy from the Sun, hit the surface of the PV cell they are absorbed by the p-type material. The additional energy provided by these photons allows electrons to overcome the bonds holding them and move within the semiconductor material, thus creating a potential difference (generating a voltage).

Photovoltaic cells have an output voltage of 0.5 V, so a number of these are linked together to form modules with resulting higher voltage and power outputs. Modules are connected together in series to increase voltage. These are known as 'strings'. All the modules together are known as an 'array'. An array can comprise a single string or multiple strings. The connection arrangements are determined by the size of the system and the choice of inverter. It should be noted that PV arrays can attain d.c. voltages of many hundreds of volts.

PV cell

ACTIVITY

What is one danger associated with photovoltaic systems?

There are many arrangements for PV systems but they can be divided into two categories:

- off-grid systems, where the PV modules are used to charge batteries
- on-grid systems, where the PV modules are connected to the grid supply via an inverter.

The key components of an off-grid PV system are:

- PV modules
- a PV module mounting system
- d.c. cabling
- a charge controller
- a deep-discharge battery bank
- an inverter.

Other components, such as isolators, will also be required.

Off-grid system components

Off-grid systems are ideal where no mains supply exists and there is a relatively small demand for power. Deep-discharge batteries are expensive and will need replacing within 5–10 years, depending on use.

On-grid systems where the PV modules are connected to the grid supply via an inverter

The key components of an on-grid PV system are:

- PV modules
- a PV module mounting system
- d.c. cabling
- an inverter
- a.c. cabling
- metering
- a connection to the grid.

Other components such as isolators will also be required.

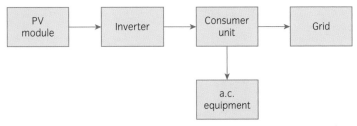

On-grid system components

PV modules

A range of different types of module, of various efficiencies, is available. The performance of a PV module is expressed as an *efficiency percentage*: the higher the percentage the greater the efficiency.

- Monocrystalline modules range in efficiency, from 15% to 20%.
- Polycrystalline modules range in efficiency, from 13% to 16%, but are cheaper to purchase than monocrystalline modules.
- Amorphous film ranges in efficiency, from 5% to 7%. Amorphous film is low efficiency but is flexible, so it can be formed into curves and is ideal for surfaces that are not flat.

Whilst efficiencies may appear low, the maximum theoretical efficiency that can be obtained with a single junction silicon cell is only 34%.

PV module mounting system

Photovoltaic modules can be fitted as on-roof systems, in-roof systems or ground-mount systems.

- On-roof systems are the common method employed for retrofit systems. Various different mounting systems exist for securing the modules to the roof structure. Most consist of aluminium rails, which are fixed to the roof structure by means of roof hooks. Mounting systems also exist for fitting PV modules to flat roofs. Checks will need to be made to ensure that the existing roof structure can withstand the additional weight and also the uplift forces that will be exerted on the PV array by the wind.

- In in-roof systems the modules replace the roof tiles. The modules used are specially designed to interlock, to ensure that the roof structure is watertight. The modules are fixed directly to the roof structure. Several different systems are on the market, from single-tile size to large panels that replace a whole section of roof tiles. In-roof systems cost more than on-roof systems but are more aesthetically pleasing. In-roof systems are generally only suitable for new-build projects or where the roof is to be retiled.

On-roof mounting system

ACTIVITY

With regards to a PV installation, who would normally be responsible for

a) mounting the roof brackets and panels

b) testing and connecting the electrical system?

In-roof mounting system

■ Ground-mount systems and pole-mount systems are available for free-standing PV arrays.

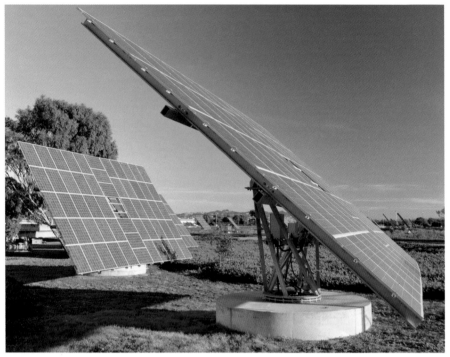

Ground-mount system

Tracking systems are the ultimate in PV mounting systems. They are computer-controlled motorised mounting systems that change both **azimuth** and tilt to track the Sun as it passes across the sky.

Inverter

The inverter's primary function is to convert the d.c. input to a 230 V a.c. 50 Hz output, and synchronise it with the mains supply frequency. The inverter also ensures that, in the event of mains supply failure, the PV system does not create a danger by continuing to feed power onto the grid. The inverter must be matched to the PV array with regard to power and d.c. input voltage, to avoid damage to the inverter and to ensure that it works efficiently. Both d.c. and a.c. isolators will be fitted to the inverter, to allow it to be isolated for maintenance purposes.

Metering

A generation meter is installed on the system to record the number of units generated, so that the feed in tariff can be claimed.

Connection to the grid

Connection to the grid within domestic premises is made via a spare way in the consumer unit and a 16 A overcurrent protective device. An isolator is fitted at the intake position to provide emergency switching, so that the PV system can easily be isolated from the grid.

Azimuth

Ideally the modules should face due south, but any direction between east and west will give acceptable outputs.

Azimuth refers to the angle by which the panel direction diverges from facing due south

PV inverter

Location and building requirements

When deciding on the suitability of a location or building for the installation of PV, the following considerations should be taken into account.

Adequate roof space available

The roof space available determines the maximum size of PV array that can be installed. In the UK, all calculations are based on 1000 Wp (watts peak) of the Sun's radiation on $1\,m^2$ so, if the array uses modules with a 15% efficiency, each 1 kWp of array will require approximately $7\,m^2$ of roof space. The greater the efficiency of the modules, the less roof space that is required.

Ideal orientation is south

The orientation (azimuth) of the PV array

The optimum direction for the solar collectors to face is due south; however, as the Sun rises in the east and sets in the west, any location with a roof facing east, south or west is suitable for mounting a PV array, but the efficiency of the system will be reduced for any system not facing due south.

The tilt of the PV array

Throughout the year, the maximum height of the Sun relative to the horizon changes. It is lowest in December and highest in June. As it is generally not practical to vary the tilt angle throughout the year, the optimum tilt for the PV array in the UK is between 30° and 40°; however, the modules will work outside the optimum tilt range and will even work if vertical or horizontal, but they will be less efficient.

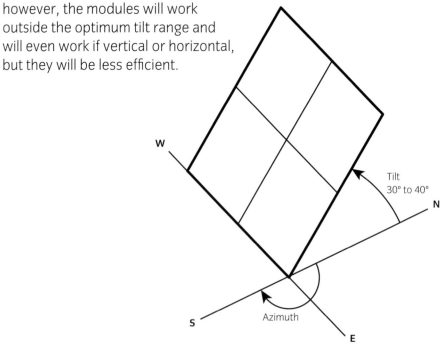

PV array facing south at fixed tilt

Ideal tilt is 30–40°

Shading of the PV array

Any structure, tree, chimney, aerial or other object that stands between the PV array (collector) and the Sun will prevent some of the Sun's energy from reaching the collector. The Sun shines for a limited time and any reduction in the amount of sunlight landing on the collector will reduces its ability to produce electricity.

Location within the UK

The location within the UK will determine how much sunshine will fall, annually, on the PV array and, in turn, this will determine the amount of electricity that can be generated. For example, a location in Brighton will generate more electricity than one in Newcastle, purely because Brighton receives more sunshine.

Shading of PV system

The suitability of the structure for mounting the solar collector

The structure has to be assessed for its suitability for fixing the chosen mounting system. Consideration needs to be given to the strength of the structure, the suitability of fixings and the condition of the structure. Consideration also needs to be given to the effect known as 'wind uplift', an upward force exerted by the wind on the module and mounting system. The strength of the PV array fixings and the fixings holding the roof members to the building structure must be great enough to allow for wind uplift.

In the case of roof-mounted systems on flats and other shared properties, consideration must also be given to the ownership of the structure on which the proposed system is to be installed.

A suitable place to mount the inverter

The inverter is usually mounted either in the loft space or at the mains position.

Connection to the grid

A spare way within the consumer unit will need to be available for connection of the PV system. If one is not available then the consumer unit may need to be changed.

Planning permission

Permitted development applies where a PV system is installed:

- on a dwelling house or block of flats
- on a building within the grounds of a dwelling house or block of flats
- as a stand-alone system in the grounds of a dwelling house or block of flats.

However, there are criteria to be met in each case.

For building-mounted systems:

- the PV system must not protrude more than 200 mm from the wall or the roof slope
- the PV system must not protrude past the highest point of the roof (the ridgeline), excluding the chimney.

For stand-alone systems the following criteria must be met.

- Only one stand-alone system is allowed in the grounds.
- The array must not exceed 4 m in height.
- The array must not be installed within 5 m of the boundary of the grounds.
- The array must not exceed 9 m^2 in area.
- No dimension of the array may exceed 3 m in length.

For both stand-alone and building-mounted systems the following criteria must be met.

- The system must not be installed in the grounds or on a building within the grounds of a listed building or a scheduled monument.
- If the dwelling is in a conservation area or a World Heritage Site, then the array must not be closer to a highway than the house or block of flats.
- In every other case, planning permission will be required.

Compliance with building regulations

The following building regulations will apply.

Part	Title	Relevance
A	Structural safety	The PV modules will impose both downward force and wind uplift stresses on the roof structure.
B	Fire safety	The passage of cables through the building fabric could reduce the fire-resisting integrity of the structure.
C	Resistance to contaminants and moisture	The fixing brackets for on-roof systems and the passage of cables through the building fabric could reduce the moisture-resisting integrity of the structure.
E	Resistance to sound	The passage of cables through the building fabric could reduce the sound-resisting properties of the structure.
L	Conservation of fuel and power	The efficiency of the system and the building overall
P	Electrical safety	The installation of the components and wiring system

ACTIVITY

What documentation should be completed by the electrical installer after testing the new PV installation?

Other regulatory requirements

- BS 7671 Wiring Regulations will apply to the PV installation.
- G83 requirements will apply to on-grid systems up to 3.68 kW per phase; above this size the requirements of G59 will need to be complied with. These documents are published by the Energy Networks Association (ENA).
- Micro Generation Certification Scheme requirements will apply.

Advantages

- They can be fitted to most buildings.
- There is a feed-in tariff available for electricity generated, regardless of whether it is used on site or exported to the grid.
- Excess electricity can be sold back to the distribution network operator (DNO).
- There is a reduction in electricity imported.
- It uses zero carbon technology.
- It improves energy performance certificate ratings.
- There is a reasonable payback period on the initial investment.

Assessment criteria

3.2 State the application and limitations of solar photovoltaic systems

Disadvantages

- Initial cost is high.

- The system size is dependent on available, suitable roof area.

- It requires a relatively large array to offset installation costs.

- It gives variable output that is dependent on the amount of sunshine available. Lowest output is at times of greatest requirement, such as at night and in the winter. Savings need to be considered over the whole year.

- There is an aesthetic impact (on the appearance of the building).

MICRO-WIND

Wind turbines harness energy from the wind and turn it into electricity. The UK is an ideal location for the installation of wind turbines, as about 40% of Europe's wind energy passes over the UK. A micro-wind turbine installed on a suitable site could easily generate more power than would be consumed on site.

Working principles

The wind passing the rotor blades of a turbine causes it to turn. The hub is connected by a low-speed shaft to a gearbox. The gearbox output is connected to a high-speed shaft that drives a generator which, in turn, produces electricity. Turbines are available as either horizontal-axis wind turbines (HAWT) or vertical-axis wind turbines (VAWT).

A HAWT has a tailfin to turn the turbine so that it is facing in the correct direction to make the most of the available wind. The gearbox and generator will also be mounted in the horizontal plane.

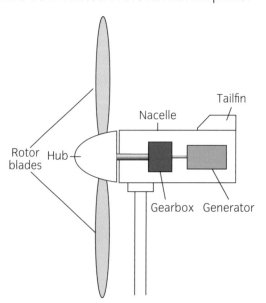

Horizontal-axis wind turbine

Vertical-axis wind turbines, of which there are many different designs, will work with wind blowing from any direction and therefore do not require a tailfin. A VAWT also has a gearbox and generator.

The two types of micro-wind turbines suitable for domestic installation are:

- pole-mounted, freestanding wind turbines
- building-mounted wind turbines, which are generally smaller than pole-mounted turbines.

Micro-wind generation systems fall into two basic categories:

- on-grid (grid tied), which is connected in parallel with the grid supply via an inverter
- off-grid, which charge batteries to store electricity for later use.

The output from a micro-wind turbine is *wild* alternating current (a.c.). 'Wild' refers to the fact that the output varies in both voltage and frequency.

The output is connected to a system controller, which rectifies the output to d.c.

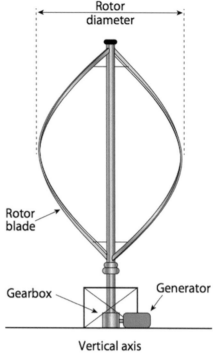

Vertical-axis wind turbine

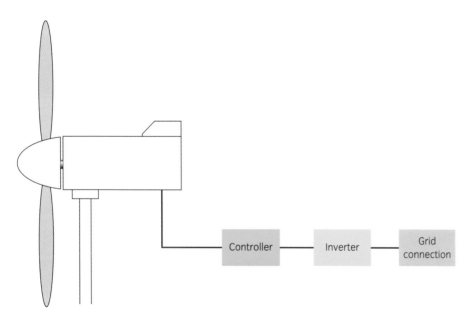

Block diagram of an on-grid micro-wind system

In the case of an on-grid system the d.c. output from the system controller is connected to an inverter which converts d.c. to a.c. at 230 V 50 Hz, for connection to the grid supply via a generation meter and the consumer unit.

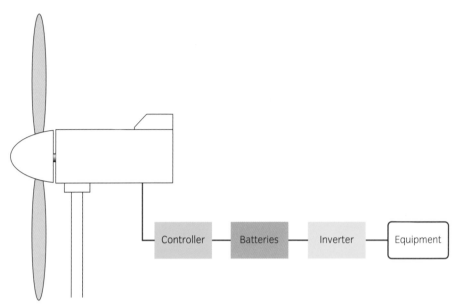

Block diagram of an off-grid micro-wind system

With off-grid systems the output from the controller is used to charge batteries so that the output is stored for when it is needed.

The output from the batteries then feeds an inverter so that 230 V a.c. equipment can be connected.

Location and building requirements

When considering the installation of a micro-wind turbine, it is important to consider the location or building requirements, including:

- the average wind speed on the site
- any obstructions and turbulence
- the height at which the turbine can be mounted
- turbine noise, vibration, flicker.

Wind speed

Wind is not constant, so the average wind speed on a site, measured in metres per second (m/s), is a prime consideration when deciding on a location's suitability for the installation of a micro-wind turbine.

Wind speed needs to be a minimum of 5 m/s for a wind turbine to generate electricity. Manufacturers of wind turbines provide power curves for their turbines, which show the output of the turbine at different wind speeds. Most micro-wind turbines will achieve their maximum output when the wind speed is around 10 m/s.

Obstructions and turbulence

For a wind turbine to work efficiently, a smooth flow of air needs to pass across the turbine blades.

The ideal site for a wind turbine would be at the top of a gentle slope. As the wind passes up the slope it gains speed, resulting in a higher output from the turbine.

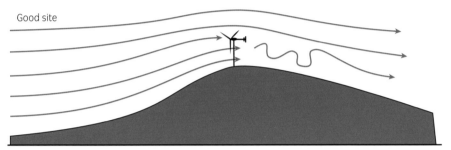

A suitable site for a micro-wind turbine

The diagram below illustrates the effect on the wind when a wind turbine is poorly sited. The wind passing over the turbine blades is disturbed and thus the efficiency is reduced.

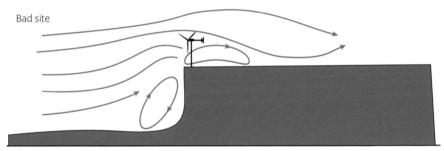

An unsuitable site for a micro-wind turbine

Any obstacles, such as trees or tall buildings, will affect the wind passing over the turbine blades.

Where an obstacle is upwind of the wind turbine, in the direction of the prevailing wind, the wind turbine should be sited at a minimum distance of 10 × the height of the obstacle away from the obstacle. In the case of an obstacle that is 10 m in height, this would mean that the wind turbine should be sited a minimum of 10 × 10 m away, which is 100 m from the obstacle.

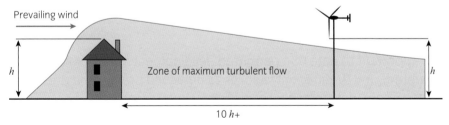

Placement of micro-wind turbine to avoid obstacles

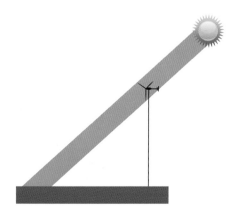
The area affected by shadow flicker

Turbine mounting height

Generally, the higher a wind turbine is mounted, the better. The minimum recommended height is 6–7 m but, ideally, it should be mounted at a height of 9–12 m. Where an obstacle lies upwind of the turbine, the bottom edge of the blade should be above the height of the obstacle. In addition, as a wind turbine has moving parts, consideration needs to be given to access for maintenance.

Turbine noise

Consideration needs to be given to buildings sited close to the wind turbine as the turbine will generate noise in use.

Turbine vibration

Consideration needs to be given to vibration when the wind turbine is building mounted. It may be necessary to consult a structural engineer.

Shadow flicker

Shadow flicker is the result of the rotating blades of a turbine passing between a viewer and the Sun. It is important to ensure that shadow flicker does not unduly affect a building sited in the shadow-flicker zone of the wind turbine.

The distance of the shadow-flicker zone from the turbine will be at its greatest when the Sun is at its lowest in the sky.

Planning permission

Whilst permitted development exists for the installation of wind turbines, it is severely restricted so, in the majority of installations, a planning application will be required. The permitted development criteria are detailed below.

Permitted development applies where a wind turbine is installed:

- on a detached dwelling house
- on a detached building within the grounds of a dwelling house or block of flats
- as a stand-alone system in the grounds of a dwelling house or block of flats.

It is important to note that permitted development for building-mounted wind turbines only applies to detached premises. It does not apply to semi-detached houses or flats.

Even with detached buildings or stand-alone turbines there are criteria to be met.

■ The wind turbine must comply with the MCS planning standards, or equivalent.

■ Only one wind turbine may be installed on the building or within the grounds of the building.

■ An air source heat pump may not be installed on the building or within the grounds of the building.

■ The highest part of the wind turbine (normally the blades) must not protrude more than 3 m above the ridge line of the building or be more than 15 m in height.

■ The lowest part of the blades of the wind turbine must be a minimum of 5 m from ground level.

■ The wind turbine must be a minimum of 5 m from the boundary of the premises.

■ The wind turbine cannot be installed on or within:

  □ land that is safeguarded land (usually designated for military or aeronautical reasons)

  □ a site that is designated as a scheduled monument

  □ a listed building

  □ the grounds of a listed building

  □ land within a national park

  □ an area of outstanding natural beauty

  □ the Broads (wetlands and inland waterways in Norfolk and Suffolk).

■ The wind turbine cannot be installed on the roof or wall of a building that fronts a highway, if that building is within a conservation area.

The following conditions also apply.

■ The blades must be made of non-reflective material.

■ The wind turbine should be sited so as to minimise its effect on the external appearance of the building.

ACTIVITY

It is possible that wind turbines may have an adverse effect on wildlife. Identify which group could be affected.

Compliance with building regulations

The following building regulations will apply.

Part	Title	Relevance
A	Structural safety	A wind turbine mounted on a building will exert additional structural load, as well as forces, due to its operation.
B	Fire safety	Cable entries and fixings may reduce the fire-resisting integrity of the building structure.
C	Resistance to contaminants and moisture	Cable entries and fixings may reduce the moisture-resisting integrity of the building fabric.
E	Resistance to sound	Cable entries may reduce the sound-resisting integrity of the building fabric.
L	Conservation of fuel and power	The efficiency of the system and the building
P	Electrical safety	Installation of wiring and components

Other regulatory requirements

- For on-grid systems, the requirements of the Distribution Network Operator (DNO) will apply together with G83 and G59, as published by the ENA.

- Wiring Regulations BS 7671 will apply to the installation of micro-wind turbines.

Assessment criteria

3.2 State the applications and limitations of micro-wind systems

Advantages

- They can be very effective on a suitable site as the UK has 40% of Europe's wind resources.

- There are no carbon dioxide emissions.

- They produce most energy in winter, when consumer demand is at its maximum.

- A feed-in tariff is available.

- This can be a very effective technology where mains electricity does not exist.

ACTIVITY

As an incentive to home owners and businesses to install PV systems, the government pays for each unit of electricity delivered into the grid. By what name is this payment known?

Disadvantages

- Initial costs are high.
- The requirements of the site are onerous.
- Planning can be onerous.
- Performance is variable and is dependent on wind availability.
- Micro-wind turbines cause noise, vibration and flicker.

MICRO-HYDRO-ELECTRIC

All rivers flow downhill. This movement of water from a higher level to a lower level is a source of free kinetic energy that hydro-electric generation harnesses. Water passing across or through a turbine can be used to turn a generator and thus produce electricity. Given the right location, micro-hydro-electric is the most constant and reliable source of all the micro-generation technologies and is the most likely of the technologies to meet all of the energy needs of the consumer.

As with the other micro-generation technologies, there are two possible system arrangements for micro-hydro schemes: on-grid and off-grid systems.

Working principles

Whilst it is possible to place generators directly into the water stream, it is more likely that the water will be diverted from the main stream or river, through the turbine, and back into the stream or river at a lower level. Apart from the work involved with the turbines and generators, there is also a large amount of civil engineering and construction work to be carried out to route the water to where it is needed.

The main components of the water course construction are:

- intake – the point where a portion of the river's water is diverted from the main stream
- the canal that connects the intake to the forebay
- the forebay, which holds a reservoir of water that ensures that the penstock is pressurised at all times and allows surges in demand to be catered for
- the penstock, which is pipework taking water from the forebay to the turbines
- the powerhouse, which is the building housing the turbine and the generator
- the tailrace, which is the outlet that takes the water exiting the turbines and returns it to the main stream of the river.

See the diagram on the next page.

Assessment criteria

3.1 Describe the fundamental operating principles of micro-hydro systems

ACTIVITY

What types of fish are most likely to be adversely affected by the installation of turbines in rivers?

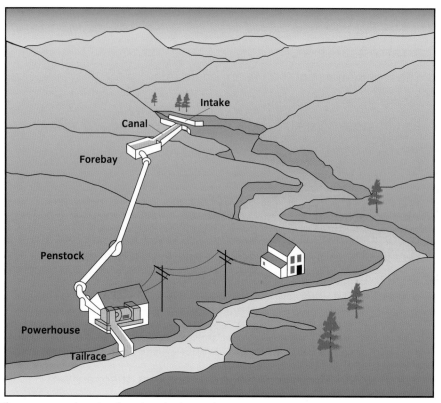

The component parts of a micro-hydro system

To ascertain the suitability of the water source for hydro-electric generation, it is necessary to consider the head and the flow of the water source.

Head

The head is the vertical height difference between the proposed inlet position and the proposed outlet. This measurement is known as 'gross head'.

Head height is generally classified as:

- low head – below 10 m
- medium head – 10–50 m
- high head – above 50 m.

There is no absolute definition for each classification. The Environment Agency, for example, classifies low head as below 4 m. Some manufacturers specify high head as above 300 m.

Net head

This is used in calculations of potential power generation and takes into account losses due to friction, as the water passes through the penstock.

Flow

This is the amount of water flowing through the water course and is measured in cubic metres per second (m³/s).

The meaning of 'head' and 'flow'

Turbines

There are many different types of turbine but they fall into two primary design groups, each of which is better suited to a particular type of water supply.

Impulse turbine

In an impulse turbine, the turbine wheel or *runner* operates in air, with water jets driving the runner. The water from the penstock is focused on the blades by means of a nozzle. The velocity of the water is increased but the water pressure remains the same so there is no requirement to enclose the runner in a pressure casing. Impulse turbines are used with high-head water sources.

Examples of impulse turbines are described below.

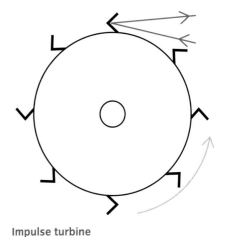

Impulse turbine

Pelton

This consists of wheel with bucket-type vanes set around the rim. The water jet hits the vane and turns the runner. The water gives up most of its energy and falls into a discharge channel below. A multi-jet Pelton turbine is also available. This type of turbine is used with water sources with medium or high heads of water.

Turgo

This is similar to the Pelton but the water jet is designed to hit the runner at an angle, from one side of the turbine. The water enters at one side of the runner and exits at the other, allowing the Turgo turbine to be smaller than the Pelton for the same power output. This type of turbine is used with water sources with medium or high heads of water.

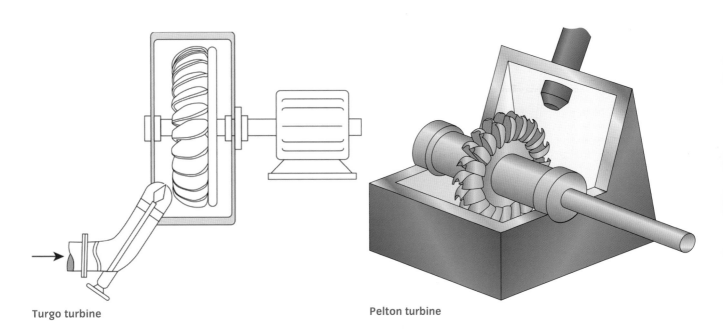

Turgo turbine

Pelton turbine

Cross-flow or Banki

With this type of turbine the runner consists of two end-plates with slats, set at an angle, joining the two discs, much like a water wheel. Water passes through the slats, turning the runner and then exiting from below. This type of turbine is used with water sources with low or medium heads of water.

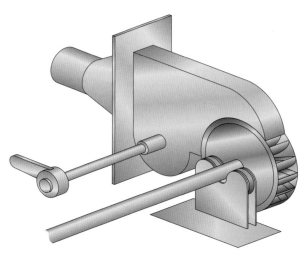

Cross-flow or Banki turbine

Reaction turbine

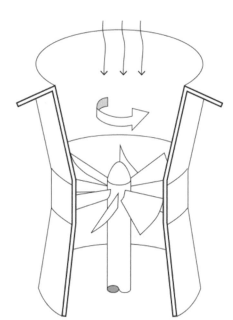

Reaction turbine

In the reaction turbine, the runners are fully immersed in water and are enclosed in a pressure casing. Water passes through the turbine, causing the runner blades to turn or react.

Examples of reaction turbines are described below.

Francis wheel

Water enters the turbine housing and passes through the runner, causing it to turn. This type of turbine is used with water sources with low heads of water.

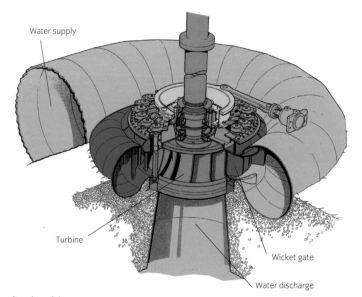

Water supply

Turbine

Wicket gate

Water discharge

Francis wheel turbine

Kaplan (propeller)

This works like a boat propeller in reverse. Water passing the angled blades turns the runner. This type of turbine is used with water sources with low heads of water.

Reverse Archimedes' screw

The Archimedes' screw consists of a helical screw thread, which was originally designed so that turning the screw – usually by hand – would draw water up the thread to a higher level. In the case of hydro-electric turbines, water flows down the screw, hence *reverse*, turning the screw, which is connected to the generator. This type of turbine is particularly suited to low-head operations but its major feature is that, due to its design, it is 'fish-friendly' and fish are able to pass through it, so it may be the only option if a hydro-electric generator is to be fitted on a river that is environmentally sensitive.

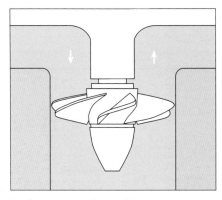

Kaplan or propeller turbine

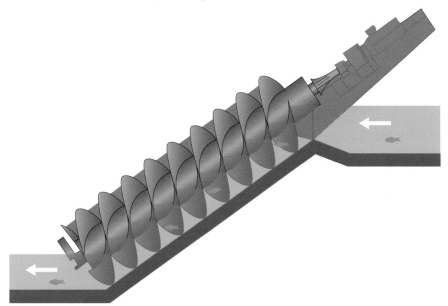

Reverse Archimedes' screw

Location building requirements

When considering the installation of a micro-hydro turbine, the following location or building requirements should be taken into account.

- The location will require a suitable water source with:
  - □ a minimum head of 1.5 m
  - □ a minimum flow rate of 100 litres/second.

The water source should not be subject to seasonal variation that will take the water supply outside of the above parameters.

- The location has to be suitable to allow construction of:
  - □ the water inlet
  - □ the turbine/generator building
  - □ the water outlet or tailrace.

Assessment criteria

3.2 State the applications and limitations of micro-hydro systems

ASSESSMENT GUIDANCE

You should be able to draw the basic layout of a micro-hydro system.

Assessment criteria

3.3 State the Local Authority Building Control requirements which apply to the installation of micro-hydro systems

Planning permission

Planning permission will be required.

A micro-hydro scheme will have an impact on:

- the landscape and visual amenity
- nature conservation
- the water regime.

The planning application will need to be accompanied by an environmental statement detailing any environmental impact and what measures will be taken to minimise these. The environmental statement typically covers:

- flora and fauna
- noise levels
- traffic
- land use
- archaeology
- recreation
- landscape
- air and water quality.

Compliance with building regulations

The following building regulations will apply.

Part	Title	Relevance
A	Structural safety	If any part of the system is housed in or connected to the building, then structural considerations will need to be taken into account.
B	Fire safety	Where cables pass through the building fabric they may reduce the fire-resisting properties of the building fabric.
C	Resistance to contaminants and moisture	Where cables pass through the building fabric they may reduce the moisture-resisting properties of the building fabric.
E	Resistance to sound	Where cables pass through the building fabric they may reduce the sound-resisting properties of the building fabric.
P	Electrical safety	Installation of components and cables

ACTIVITY

What are the requirements of BS 7671 regarding repairing holes made in walls for the passage of cables?

Other regulatory requirements

- BS 7671 Wiring Regulations
- G83 requirements for grid-tied systems
- Micro Generation Certification Scheme requirements
- Environment Agency requirements

In England and Wales, all waterways of any size are controlled by the Environment Agency.

To remove water from these waterways, even though it may be returned – as in the case of a hydro-electric system – will usually require permission and a licence.

There are three types of licence that may apply to a hydro-electric system.

- An *abstraction licence* will be required if water is diverted away from the main water course. The major concern will be the impact that the project has on fish migration, as the majority of turbines are not fish-friendly.

  This requirement may affect the choice of turbine (*see* Reverse Archimedes' screw, page 147). It may mean that fish screens are required over water inlets or, where the turbine is in the main channel of water, a fish pass around the turbine may need to be constructed.

- An *impoundment licence* – an impoundment is any construction that changes the flow of water, so if changes or additions are made to sluices, weirs, etc that control the flow within the main stream of water, an impoundment licence will be required.

- A *land drainage licence* will be required for any changes made to the main channel of water.

An Environment Site Audit (ESA) will be required as part of the initial assessment process. The ESA covers:

- water resources
- conservation
- chemical and physical water quality
- biological water quality
- fisheries
- managing flood risk
- navigation of the waterway.

Assessment criteria

3.2 State the applications and limitations of micro-hydro systems

Advantages

- There are no on-site carbon emissions.
- Large amounts of electricity are output, usually more than required for a single dwelling. The surplus can be sold.
- A feed-in tariff is available.
- There is a reasonable payback period.
- It is an excellent system where no mains electricity exists.
- It is not dependent on weather conditions or building orientation.

Disadvantages

- It requires a high head or fast flow of water on the property.
- It requires planning permission, which can be onerous.
- Environment Agency permission is required for water extraction.
- It may require strengthening of the grid for grid-tied systems.
- Initial costs are high.

Applying co-generation energy technologies

CO-GENERATION TECHNOLOGIES – MICRO-COMBINED HEAT AND POWER (HEAT-LED)

In micro-combined heat and power (mCHP) technology, a fuel source is used to satisfy the demand for heat but, at the same time, generates electricity that can either be used or sold back to the supplier.

Currently, mCHP units used in domestic dwellings are powered by means of natural gas or liquid propane gas (LPG), but could be fuelled by using biomass, liquid propane gas (LPG) or other fuels.

The diagram below represents, from left to right, an old, inefficient gas boiler, a modern condensing boiler and an mCHP unit.

Baxi Ecogen domestic CHP boiler

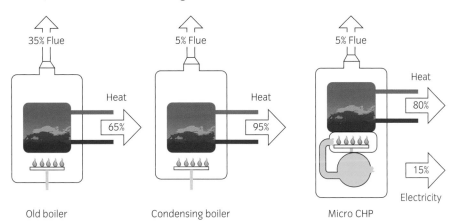

The efficiency of different boilers

With the old, inefficient boiler, 65% of the input energy is used to provide heating for the premises, 35% is lost up the flue. With the condensing boiler, this lost heat is re-used so that the output to the heating is 95%. The mCHP unit will achieve the same efficiencies as the modern condensing boiler but 80% of the input is used to provide heat and 15% is used to power a generator to provide power.

There are obvious savings to be made in replacing an old, inefficient boiler with a mCHP unit, but could the same savings not be made by fitting a condensing boiler? At first sight this may appear to be feasible; however, on a unit-by-unit comparison, gas is cheaper than electricity, so any electricity generated by using gas means a proportionally greater financial saving over using electricity.

ASSESSMENT GUIDANCE

The feed-in tariff for mCHP is slightly less than the PV rate and considerably less than wind and hydro tariffs at 2013 rates. For latest rates visit the Energy Saving Trust: www.energysavingtrust.org.uk.

Assessment criteria

3.1 Describe the fundamental operating principles of micro-combined heat and power (heat-led) systems

ACTIVITY

Find out an alternative name for a Stirling engine.

In addition to this saving, locally generated power reduces transmission losses and consequently creates less carbon dioxide (CO_2) than if the electricity were generated at a power station some distance away.

This type of generation, using a mCHP unit, is known as 'heat-led', as the primary function of the unit is to provide space heating, while the generation of electricity is secondary. The more heat that is produced, the more electricity is generated. The unit only generates electricity when there is a demand for heating. Most domestic mCHP units will generate between 1 kW and 1.5 kW of electricity. Micro-combined heat and power is a carbon-reduction technology rather than a carbon-free technology.

Working principles

Combined heat and power (CHP) units have been available for a number of years but it is only recently that domestic versions have become available. Domestic versions are usually gas-fired and use a Stirling engine to produce electricity, though other fuel sources, and types of generator combinations, are available.

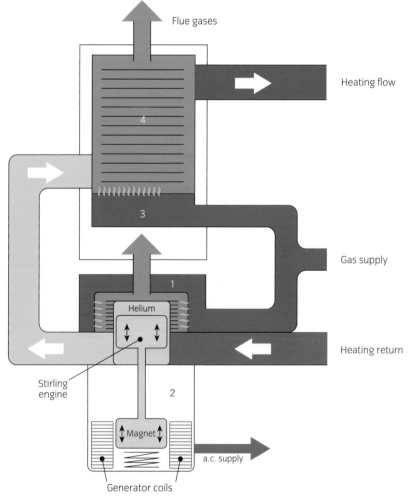

Component parts of a micro-CHP boiler

The key components of the mCHP unit are:

1 the engine burner

2 the Stirling engine generator

3 the supplementary burner

4 the heat exchanger.

When there is a call for heat, the engine burner fires and starts the Stirling engine generator. The engine burner produces about 25% of the full heat output of the unit. The burner preheats the heating-system return water before passing it to the main heat exchanger. The hot flue gases from the engine burner are passed across the heat exchanger to heat the heating-system water further.

If there is greater demand than is being supplied by the engine burner, then the supplementary burner operates to meet this demand.

How the Stirling engine generator works

The first Stirling engine was invented by Robert Stirling in 1816 and is very different from the internal combustion engine.

The Stirling engine uses the expansion and contraction of internal gases, due to changes in temperature, to drive a piston. The gases within the engine do not leave the engine and no explosive combustion takes place, so the Stirling engine is very quiet in use.

In the case of the Stirling engine used in an mCHP unit, the gas contained within the engine is helium.

When the engine burner fires, the helium expands, forcing the piston downwards. The return water from the heating system passing across the engine cools the gas, causing it to contract. A spring arrangement within the engine returns the piston to the top of the cylinder and the process starts all over again.

The piston is used to drive a magnet up and down between coils of wire, generating an electromotive force (emf) in the coils.

Connection of the mCHP unit to the supply

The preferred connection method between the mCHP unit and the electricity supply is via a dedicated circuit, directly from the consumer unit. This method will allow for easy isolation of the generator from the incoming supply.

Assessment criteria

3.2 State the applications and limitations of micro-combined heat and power (heat-led) systems

Location and building requirements

For mCHP to be viable, the following criteria must be met.

- The building should have a high demand for space heating. The larger the property the greater the carbon savings.
- A building that is well insulated will not usually be suitable, as a well-insulated building is unlikely to have a high demand for space heating.

If an mCHP unit is fitted to a building that is either too small or is well insulated, this will mean that the demand for heat will be small and the mCHP unit will cycle on and off, resulting in inefficient operation.

KEY POINT

Micro-CHP boilers are only suitable where there is a high demand for heat.

Assessment criteria

3.3 State the Local Authority Building Control requirements which apply to the installation of micro-combined heat and power (heat-led) systems

Planning permission

Planning permission will not normally be required for the installation of an mCHP unit in a domestic dwelling if all of the work is internal to the building.

If the installation requires an external flue to be installed, this will normally be classed as permitted development as long as the following criterion is met.

- Flues to the rear or side elevation do not extend more than 1 m above the highest part of the roof.

Listed building or buildings in a designated area

Check with the local planning authority regarding both internal work and external flues.

Buildings in a conservation area or in a World Heritage Site

Flues should not be fitted on the principle or side elevation if they would be visible from a highway.

ASSESSMENT GUIDANCE

It is necessary to consider not only the purchase costs but also the running and maintenance costs of any system.

If the project includes the construction of buildings for fuel storage, or to house the mCHP unit then the same planning requirements as for extensions and garden outbuildings will apply.

Compliance with building regulations

The following building regulations will apply.

Part	Title	Relevance
A	Structural safety	Where the components increase load on the structure or where holes reduce the structural integrity of the building
B	Fire safety	Where installation of the system decreases the fire-integrity of the structure, for example, where pipes or cables pass through fire compartments
C	Resistance to contaminants and moisture	Holes for pipes or cables could reduce the moisture-resisting integrity of the building.
E	Resistance to sound	Holes for pipes or cables could reduce the sound-resisting qualities of the building.
G	Sanitation, hot-water safety, and water efficiency	Hot-water safety and water efficiency
J	Heat-producing appliances	mCHP units are heat-producing systems.
L	Conservation of fuel and power	Energy efficiency of the system and the building
P	Electrical safety	Electrical installation of controls and supply

ACTIVITY

State whether BS 7671 IET Wiring Regulations is a statutory or non-statutory document.

Other regulatory requirements

- Gas regulations will apply to the installation of the mCHP unit. The gas installation work will need to be carried out by an operative registered on the Gas Safe register.

- Water regulations (WRAS) will apply to the water systems.

- BS 7671 Wiring Regulations will apply to the installation of control wiring and the wiring associated with the connection of the mCHP electrical generation output.

- G83 requirements will apply to the connection of the generator, although mCHP units do have a number of exemptions.

- Micro Generation Certification Scheme requirements.

Assessment criteria

3.2 State the applications and limitations of micro-combined heat and power (heat-led) systems

Advantages

- The ability to generate electricity is not dependent on building direction or weather conditions.

- The system generates electricity whilst there is a need for heat.

- A feed-in tariff is available but is limited to generator outputs of less than 2 kW and is only applicable to the first 30 000 units.

- Saves carbon over centrally generated electricity.

- Reduces the building's carbon footprint.

Disadvantages

- The initial cost of an mCHP is high, when compared to an efficient gas boiler.

- It is not suitable for properties with low demand for heat – small or very well insulated properties.

- There is limited capacity for generation of electricity.

Applying water conservation technologies

Many people regard the climate of the United Kingdom as wet. It is a common perception that the UK has a lot of rain and, in some locations, this is true, especially towards the west, where average annual rainfall is in excess of 1000 mm, but along the east coast the average is less than half of this.

The population of the UK is expanding and the demands on the water supply systems are ever increasing. Hose-pipe bans in many parts of the country are a regular feature of the summer months. In the UK, unlike many of our European neighbours, the water supplied is suitable for consumption straight from the tap, but we use it not only for drinking, but for bathing, washing clothes, watering gardens and washing cars.

Even in the UK, clean, fresh water is a limited resource. With growing demand, the pressure on this vital resource is increasing. Water conservation is one way of ensuring that demand does not outstrip supply and that shortages are avoided.

The two methods of water conservation covered in this unit are:

- rainwater harvesting
- re-use of greywater.

Water conservation is one way of reducing water bills. Whether calculated as measured (metered) or unmeasured, water bills, both contain two charges:

- charges for fresh water supplied
- charges for sewage or waste water taken away.

The amount of water taken away, which includes surface water (rainwater), is assumed to be 95% of the water supplied.

By conserving water, the waste-water charge, as well as the charge for fresh water supplied, can be reduced. Besides producing a reduction in household bills, water conservation will also help to relieve the pressure on a vital resource.

ACTIVITY

Water conservation has been used for hundreds of years, especially by gardeners. How was this achieved?

The Code for Sustainable Homes sets a target for reducing average drinking water consumption from 150 litres per person per day to an optimum 80 litres. The adopted target is currently 103 litres. Part G of the Building Regulations sets the level at 125 litres. Whichever target is used, the conclusion to be drawn is that a reduction in consumption is vital.

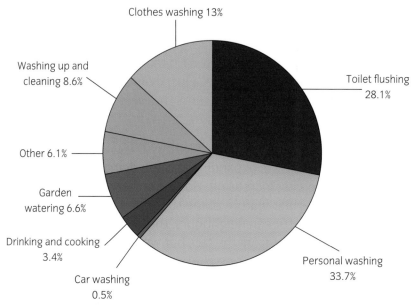

How water is used

Within the average home the amount of water used for drinking or food preparation is estimated at only 3.4% of total consumption, but there are obvious opportunities elsewhere for water savings.

The technologies to be covered are not concerned with direct carbon reduction or with financial savings, but are concerned with the reduction in consumption of a valuable resource.

Terminology used	Meaning
Wholesome water	Water that is palatable and suitable for human consumption. The water that is obtained from the utility company supply and is also known as 'wholesome', 'potable' and 'white water'
Rainwater	Water captured from rainwater gutters and downpipes
Greywater	Waste water from wash basins, showers, baths, sinks and washing machines
Black water	Sewage

RAINWATER HARVESTING

Rainwater harvesting refers to the process of capturing and storing rainwater from the surface it falls on, rather than letting it run off into drains or allowing it to evaporate. Re-use of rainwater can result in sizeable reductions in wholesome water usage and thus monetary savings as well as carbon reductions.

If harvested rainwater is filtered, stored correctly and used regularly, so that it does not remain in the storage tanks for an excessive period of time, it can be used for:

- flushing toilets
- car washing
- garden watering
- supplying a washing machine.

Harvested rainwater cannot be used for:

- drinking water
- washing of dishes
- food preparation or washing of food
- personal hygiene, ie washing, bathing or showering.

Rainwater is classified as 'fluid category 5' risk, which is the highest risk category.

Working principles

The process steps in re-using rainwater are:

- collection
- filtration
- storage
- re-use.

Collection or capture of rainwater

Rainwater can be captured from roofs or hard standings. In the case of roofs the water is captured by means of gutters and flows to the water-harvesting tank via the property's rainwater downpipes. The amount of water that can be collected will be governed by:

- the size of the capture area
- the annual rainfall in the area.

Not all of the water that falls on a surface can be captured. During periods of very heavy rainfall, water may overflow gutters or merely bounce off the roof surface and avoid the guttering system completely.

Assessment criteria

3.1 Describe the fundamental operating principles of rainwater harvesting systems

Water collected from roofs covered in asbestos, copper, lead or bitumen may not be suitable for re-use and may pose a health risk, or result in discolouration of the water or odour problems. Water collected from hard standings such as driveways may be contaminated with oil or faecal matter.

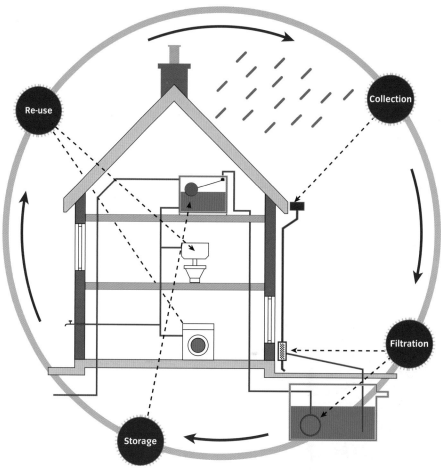

Rainwater-harvesting cycle

Filtration

As the rainwater passes from the rainwater-capture system to the storage tanks, it passes through an in-line filter to remove debris such as leaves. These are flushed out. The efficiency of this filter will determine how much of the captured water ends up in the storage tank. Manufactures usually quote figures in excess of 90% efficiency.

Storage of rainwater

Rainwater storage tanks can be either above-ground or below-ground types and can vary in size from a small tank next to a house, to a buried tank that is able to hold many thousands of litres of water. Below-ground tanks will require excavation works, whilst above-ground tanks will need a suitably sized space to be available. Whichever type of tank is used, it will need to be protected against freezing, heating from direct sunlight and contamination.

KEY POINT

Most people with a garden will have a water butt to collect rainwater. Special adaptors are available that can be cut into the drainpipe to divert the water into the water butt.

The size of the tank will be determined by the rainwater available and the annual demand. It is common practice to size a tank to be 5% of annual rainwater supply or the anticipated annual demand.

A submersible pump is used to transport water from the storage tank to the point of demand.

The tank will incorporate an overflow pipe connected to the drainage system of the property, for times when the harvested rainwater exceeds the capacity of the storage tank.

Re-use of stored rainwater

With rainwater harvesting, two system options are available for the re-use of the collected and stored water: indirect and direct distribution.

Indirect distribution system

With indirect distribution systems, water is pumped from the storage tank to a supplementary storage tank or header tank located within the premises. This, in turn, feeds the water outlets via pipework separated from the wholesome water supply pipes.

- Wholesome water inlet
- Rainwater inlet
- Air gap for water regulations compliance
- Rainwater level control
- Wholesome water level control
- Overflow
- Outlet

Header tank with backflow protection

The header tank will incorporate a backflow prevention air gap to meet the Water Regulation requirements.

The arrangement of water control and overflow pipes will ensure that the air gap is maintained. The rainwater level control connects to a control unit that operates the submersible pump, so that water is drawn from the main storage tank when required.

At times when there is not enough rainwater to meet the demand, fresh water is introduced into the system via the wholesome water inlet, which is controlled by means of the wholesome-water level control.

Direct distribution system

In direct distribution systems, the control unit pumps rainwater directly to the outlets on demand. At times of low rainwater availability, the control unit will provide water from the wholesome supply to the outlets. The backflow prevention methods to meet the requirements of the Water Regulations will be incorporated into the control unit. This type of system uses more energy than the indirect distribution system.

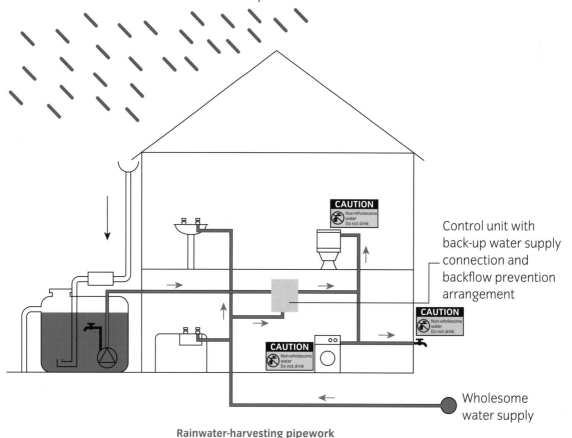

Control unit with back-up water supply connection and backflow prevention arrangement

CAUTION
Non-Wholesome water
Do not drink

CAUTION
Non-wholesome water
Do not drink

CAUTION
Non-wholesome water
Do not drink

Wholesome water supply

Rainwater-harvesting pipework

Assessment criteria

3.2 State the applications and limitations of rainwater harvesting systems

Location and building requirements

When deciding on the suitability of a location for the installation of a water-harvesting system, the following points will need to be taken into account.

■ Is there a suitable supply of rainwater to meet the demand? This is determined by finding the rainfall available and the level of water use by the occupants.

■ A suitable supply of wholesome water will be required to provide back-up at times of drought.

■ For above-ground storage tanks, the chosen location must avoid the risk of freezing, or the warming effects of sunlight, which may encourage algal growth.

■ For below-ground tanks, consideration will need to be given to access for excavation equipment.

Planning permission

In principle, planning permission is not normally required for the installation of a rainwater-harvesting system if it does not alter the outside appearance of the property. It is, however, always worth enquiring, especially if the system is installed above ground, the building is in a designated area or the building is listed.

Compliance with building regulations

The following building regulations will apply.

Assessment criteria

3.3 State the Local Authority Building Control requirements which apply to the installation of rainwater harvesting systems

Part	Title	Relevance
A	Structural safety	Where the components affect the loadings placed on the structure of the building or excavations are in close proximity to the building
B	Fire safety	Where holes for pipework reduce the fire-resisting integrity of the structure
C	Resistance to contaminants and moisture	Where holes for pipework reduce the moisture-resisting integrity of the structure, for example, pipework passing through vapour barriers
E	Resistance to sound	Where holes for pipework reduce the sound-resisting integrity of the structure
G	Sanitation, hot-water safety, and water efficiency	Water efficiency
H	Drainage and waste disposal	Where gutters and rainwater pipes are connected to the system.
P	Electrical safety	Installation of supply and control wiring for the system

Other regulatory requirements

- The Water Supply (Water Fittings) Regulations 1999 apply to rainwater-harvesting systems. The key area of concern will be the avoidance of cross-contamination between rainwater and wholesome water. This is known as 'backflow prevention' and, as rainwater is classified as a category 5 risk, the usual method of providing backflow prevention is with a type AA air-gap between the wholesome water and the rainwater.

Pipework labels

3.2 State the applications and limitations of rainwater harvesting systems

Any pipework used to supply outlets with the rainwater will need to be labelled to distinguish it from wholesome water. Outlets will also need to be labelled to indicate that the water supplied is not suitable for drinking.

Other regulatory requirements

- Wiring Regulations BS 7671 will apply to the installation of supplies and control systems for the rainwater-harvesting system.
- BS8515:2009 Rainwater Harvesting Systems – Code of Practice

Advantages

- There is a reduction in use of wholesome water.
- Water bills are reduced if the supply is metered.
- Water does not require any further treatment before use.
- The system is less complicated than greywater re-use systems.

Disadvantages

- The quantity of available water is limited by roof area. It may not meet the demand in dry periods.
- Initial costs are high.
- A water meter should be fitted.

GREYWATER RE-USE

Greywater is the waste water from baths, showers, hand basins, kitchen sinks and washing machines. It gets its name from its cloudy, grey appearance. Capturing and re-using the water for permitted uses reduces the consumption of wholesome (drinking) water.

Greywater collected from washbasins, showers and baths will often be contaminated with human intestinal bacteria and viruses, as well as organic material such as skin particles and hair. As well as these contaminants, it will also contain soap, detergents and cosmetic products, which are ideal nutrients for bacteria growth. Add to this the relatively high temperature of the greywater and ideal conditions exist to encourage the growth of bacteria.

For these reasons, untreated greywater cannot be stored for more than a few hours. The less-polluted water from washbasins, showers and baths is usually used in greywater re-use systems. This is known as 'bathroom greywater'. Where a greater supply of greywater is required, washing-machine waste water is collected.

KEY POINT

Whilst kitchen sink waste is classified as greywater, it is not usually collected and recycled. This is because the FOGs (fats, oils and greases) contained within it will emulsify as the water cools and will not be kind to the filtration systems.

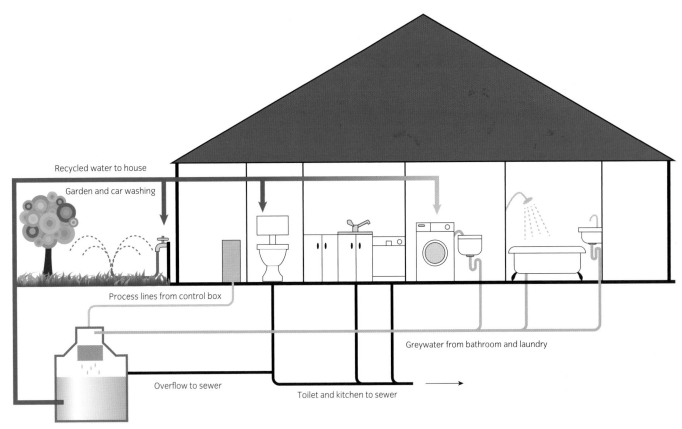

Greywater re-use system

Greywater is classified as fluid category 5 risk (the highest) under the Water Supply (Water Fittings) Regulations 1999. Greywater can pose a serious health risk, due to its potential pathogen content. Untreated greywater deteriorates rapidly when stored, so all systems that store greywater will need to incorporate an appropriate level of treatment.

If greywater is filtered and stored correctly then it can be used for:

- flushing toilets
- car washing
- garden watering
- washing clothes (after additional processing).

Greywater cannot be used for:

- drinking water
- washing of dishes
- food preparation or washing of food
- personal hygiene – washing, bathing or showering.

ACTIVITY

Currently, greywater is normally discharged directly into the drainage system of the house and then to the main drainage system. Think of the changes that would have to be made to direct all the greywater to a central collection point.

Assessment criteria

3.1 Describe the fundamental operating principles of greywater re-use systems

Ecoplay greywater system
1 Cleaning tank
2 Storage tank

ACTIVITY

Go to Water Works UK www.wwuk.co.uk or alternative websites and see the range of systems available.

Working principles

Several types of greywater re-use systems exist but, apart from the direct re-use system, they all have similar common features:

- a tank for storage of the treated water
- a pump
- a distribution system for moving the water from storage to where it is to used
- some form of treatment.

Direct re-use

Greywater is collected from appliances and directly re-used, without treatment or storage, and can be used for such things as watering the garden. Even so, the greywater is not considered suitable for watering fruit or vegetable crops.

Short retention system

Greywater from baths and showers is collected in a cleaning tank, where it is treated, by means such as surface skimming, to remove debris such as soap, hair and foam. Heavier particles are allowed to settle to the bottom, where they are flushed away as waste. The remaining water is then transferred to a storage tank, ready for use. The storage tanks are usually relatively small, at around 100 litres, which is enough for 18–20 toilet flushes. If the water is not used within a short time, generally 24 hours, the stored greywater is purged, the system is cleaned and a small amount of flush water is introduced to allow toilet flushing. This avoids the greywater deteriorating, and beginning to smell, at times when the premises are unoccupied for a lengthy period of time. This type of system can result in water savings of 30%.

This type of system would be ideal for installation in a new-build project but would be more difficult to retrofit. It is usually fitted in the same room as the source of greywater.

Physical and chemical system

This system uses a filter to remove debris from the collected water (physical cleaning). After the greywater has been filtered, chemical disinfectants such as chlorine or bromine are added, to inhibit bacterial growth during storage.

Biomechanical system

This type of system is the most advanced of the greywater re-use systems. It uses both biological and physical methods to treat the collected greywater. An example of such a system is the German system 'AquaCycle® 900', which comprises an indoor unit about the size of a large refrigerator.

Greywater enters the system and passes through the filtering unit (1), where particles such as hair and textiles debris are filtered out. The filtering unit is electronically controlled to provide automatic flushing of the filter.

Water enters the main recycling chamber (2), where organic matter is decomposed by bio-cultures. The water remains in this chamber for 3 hours before being pumped to the secondary recycling chamber (3), for further biological treatment. Biological sediment settles to the bottom of each chamber (4), where it is sucked out and transferred to a drain.

After a further period of 3 hours, the water passes through a UV filter (5), to the final storage chamber (7), where it is ready for use. When there is demand for the treated water this is pumped (8) to the point of demand. At times when treated water availability is low, fresh water (6) can be introduced to the system.

Water from this unit can be used for washing clothes as well as the other uses previously stated.

AquaCycle® 900 system

Biological system

This type of system uses some of the principles employed by sewage treatment works. In this case, bacterial growth is encouraged rather than inhibited, by the introduction of oxygen to the waste water. Oxygen can be introduced by means of pumps pushing air through the storage tanks. Bacteria then 'digest' the organic matter contained within the greywater.

A more 'natural' method of oxygenating the water is by the use of reed beds. In nature, reeds, which thrive in waterlogged conditions, transfer oxygen to their roots. The greywater is allowed to infiltrate reed beds. The added oxygen and naturally occurring bacteria will remove any organic matter contained in the waste water. The disadvantages of using reed beds are the land area required for the reed beds and the expertise required to maintain them.

Alternative Water Solutions produces a system based on these principles, but on a smaller scale, called the Green Roof Recycling System (GROW). This uses a system of tiered gravel-filled troughs planted with native plants, the roots of which can perform the same function of filtering as a reed bed would.

GROW system

Assessment criteria

3.2 State the applications and limitations of greywater re-use systems

Location and building requirements

When considering the installation of a greywater re-use system, the following location or building requirements should be taken into account.

- There needs to be a suitable supply of greywater to meet the demand. Premises with a low volume of greywater are not suitable.
- Suitability of the location and availability of space to store enough greywater to meet the demand of the premises must be assessed.
- Storage tanks need to be located away from heat, including direct sunshine, to avoid the growth of algae. They need to be located so they are not subject to freezing in cold weather. There needs to be a wholesome water supply.
- Where greywater tanks are retrofitted, access for excavation equipment will need to be considered.
- A water meter will need to be fitted on the water supply to maximise the benefits.

Assessment criteria

3.3 State the Local Authority Building Control requirements which apply to the installation of grey water re-use systems

Planning permission

In principle, planning permission is not normally required for the installation of a greywater re-use system, if the system does not alter the outside appearance of the property. It is, however, always worth enquiring, especially if the system is installed above ground, the building is in a designated area or the building is listed.

If a building is required to house the greywater storage system, then a planning application will need to be submitted.

Compliance with building regulations

The following building regulations will apply.

Part	Title	Relevance
A	Structural safety	Where the components affect the loadings placed on the structure of the building or excavations are in close proximity to the building
B	Fire safety	Where holes for pipework reduce the fire-resisting integrity of the structure
C	Resistance to contaminants and moisture	Where holes for pipework reduce the moisture-resisting integrity of the structure, for example, pipework passing through vapour barriers
E	Resistance to sound	Where holes for pipework reduce the sound-resisting integrity of the structure
G	Sanitation, hot-water safety, and water efficiency	Water efficiency
H	Drainage and waste disposal	Where waste pipes are connected to the system
P	Electrical safety	Installation of supply and control wiring for the system

Other regulatory requirements

- The Water Supply (Water Fittings) Regulations 1999 apply to greywater recycling installations. The key area of concern will be the avoidance of cross-contamination between greywater and wholesome water. This is known as 'backflow prevention' and, as greywater is classified as a category 5 risk, the usual method of providing backflow prevention is with an air gap between the wholesome water and the greywater.

- Any pipework used to supply outlets with the treated greywater will need to be labelled to distinguish it from pipework for wholesome water. Outlets will need to be clearly identified by means such as labelling. Outlets will also need to be labelled to indicate that the water supplied is not suitable for drinking.

Greywater warning label

Other regulatory requirements

■ The local water authority must be notified when a greywater re-use system is to be installed.

■ Wiring Regulations BS 7671 will apply to the installation of supplies and control systems for the greywater re-use system.

Assessment criteria

3.2 State the applications and limitations of greywater re-use systems

Advantages

■ There will be a reduction in water bills if the supply is metered.

■ It reduces demands on the wholesome water supply.

■ A wide range of system options exists.

■ It has the potential to provide more re-usable water than a rainwater harvesting system.

Disadvantages

■ There are long payback periods.

■ It can be difficult to integrate into an existing system.

■ Only certain types of appliance or outlet can be connected. This causes additional plumbing work.

■ Cross-contamination can be a problem.

■ A water meter will need to be fitted to ensure maximum gains.

■ The need for filtering and pumping may actually increase rather than decrease the carbon footprint.

KEY POINT

Traditionally, water charges were based on the rateable value of the house. Since this system was abolished, most companies kept similar charges but many are now fitting water meters.

ASSESSMENT CHECKLIST

WHAT YOU NOW KNOW/CAN DO

Learning outcome	Assessment criteria	Page number
1 Understand the environmental legislation, working practices and principles which are relevant to work activities	*The learner can:*	
	1 Specify the current, relevant legislation for processing waste	80
	2 Describe what is meant by the term environment	80
	3 Describe the ways in which the environment may be affected by work activities	86
	4 Identify and interpret the requirements for electrical installations as outlined in relevant sections of the Building Regulations and the Code for Sustainable Homes	87
	5 State materials and products that are classed as: ■ hazardous to the environment ■ recyclable	87
	6 Describe the organisational procedures for processing materials that are classed as: ■ hazardous to the environment ■ recyclable.	89
2 Understand how work methods and procedures can reduce material wastage and impact on the environment	*The learner can:*	
	1 State installation methods that can help to reduce material wastage	90
	2 Explain why it is important to report any hazards to the environment that arise from work procedures	92
	3 Specify environmentally friendly materials, products and procedures that can be used in the installation and maintenance of electrotechnical systems and equipment.	93

Learning outcome	Assessment criteria	Page number
3 Understand how and where environmental technology systems can be applied	*The learner can:*	
	1 Describe the fundamental operating principles of environmental technology systems	
	■ solar thermal (hot water)	101
	■ solar photovoltaic	128
	■ ground source heat pump	111
	■ air source heat pump	116
	■ micro-wind	136
	■ biomass	120
	■ micro-hydro	143
	■ micro-combined heat and power (heat-led)	152
	■ rainwater harvesting	159
	■ greywater re-use	166
	2 State the applications and limitations of environmental technology systems	
	■ solar thermal (hot water)	105, 108
	■ solar photovoltaic	132, 135
	■ ground source heat pump	115, 116
	■ air source heat pump	118, 120
	■ micro-wind	138, 142
	■ biomass	123, 125
	■ micro-hydro	147, 150
	■ micro-combined heat and power (heat-led)	154, 156
	■ rainwater harvesting	162, 164
	■ greywater re-use.	168, 170
	3 State the local authority building control requirements which apply to the installation of environmental technology systems.	97
	■ solar thermal (hot water)	106
	■ solar photovoltaic	134
	■ ground source heat pump	115
	■ air source heat pump	118
	■ micro-wind	140
	■ biomass	124
	■ micro-hydro	148
	■ micro-combined heat and power (heat-led)	154
	■ rainwater harvesting	163
	■ greywater re-use.	168

ASSESSMENT GUIDANCE

■ The assessment for this unit is by one e-volve on-line multiple-choice test and one written assignment on PV design and waste management.

■ For the multiple-choice test, ensure you know how the e-volve system works. Ask for a demonstration if you are not sure.

■ It is better to have time left over at the end rather than have to rush to finish.

■ Make sure you read every question carefully.

■ If you need paper ask the invigilator to provide some.

■ Make sure you have a scientific (non-programmable) calculator.

■ Do not take any paperwork into the exam with you.

■ Make sure your mobile phone is switched off during the exam. You may be asked to give it to the invigilator.

Before the assessment

■ For the e-volve test, you will find sample questions on SmartScreen and some questions in the section below to test your knowledge of the learning outcomes.

■ Make sure you go over these questions in your own time.

■ Spend time on revision in the run-up to the assessment.

OUTCOME KNOWLEDGE CHECK

1   A system that uses a fluid to capture heat from the Sun is known as:

   a)   photovoltaic

   b)   solarvoltaic

   c)   solar thermal

   d)   solar thermostatic.

2   A buffer tank can be used for storing:

   a)   cold water

   b)   rainwater

   c)   electricity

   d)   hot water.

3   A system that uses sunlight to generate electricity is known as:

a)   photovoltaic

b)   photosynthetics

c)   solar thermal

d)   solar chemical.

4   Excess electricity can be sold back to the grid using:

a)   an off-peak tariff

b)   a feed-in tariff

c)   a feed-out tariff

d)   a daytime tariff.

5   One disadvantage of ground source heat pumps using horizontal pipework is:

a)   the need for unshaded ground

b)   the need for large ground area

c)   the need for deep boreholes

d)   that it is less than 100% efficient.

6   The air source heat pump works on the principle of:

a)   cold exchange

b)   electromagnetic induction

c)   the refrigeration cycle

d)   the Otto cycle.

7   Which one of the following fuels is commonly used in biomass systems?

a)   Coal.

b)   Oil.

c)   Shale.

d)   Wood chips.

8   Rainwater harvesting should not be used for:

a)   car washing

b)   drinking water

c)   toilet flushing

d)   garden watering.

9   A micro-hydro system relies upon:

   a) a constant wind speed

   b) reliable water flow

   c) a deep well

   d) sufficient sunlight.

10 A micro-wind system generates d.c. that is changed into a.c. by the use of:

   a) a rectifier

   b) an inverter

   c) a commutator

   d) slip rings.

11 Which orientation is NOT suitable for mounting a photovoltaic panel?

   a) North.

   b) South.

   c) East.

   d) West.

12 The maximum distance a solar thermal panel may extend above the roof surface is:

   a) 50 mm

   b) 150 mm

   c) 200 mm

   d) 250 mm.

13 A ground source heat pump installed on a building with limited grounds will require:

   a) deep boreholes

   b) long pipe trenches

   c) an extra low-voltage supply

   d) shallow trenches.

14 An air source heat pump has an electrical input of 2.5 kW and output of 7.5 kW. The coefficient of performance will be:

   a) 0.33

   b) 3

   c) 5

   d) 0.

15 The position of the biomass flue should be:

   a) at the front of the building

   b) at the side or rear of a building

   c) roof-mounted at least 2 m above roof level

   d) roof-mounted at least 3 m above roof level.

16 An 'on-grid' generating system is one where the system:

   a) is connected in a 4 × 4 grid

   b) is connected to a battery bank only

   c) can be moved within an installed grid

   d) is connected to the national grid.

17 Installation of greywater systems would depend upon:

   a) total black and greywater available

   b) adequate greywater availability

   c) sufficient rainwater only

   d) direct mixing with wholesome water.

18 One disadvantage associated with biomass systems is:

   a) water produced by burning fuel

   b) inability to control heat output

   c) need for large storage area

   d) large amount of ash produced.

19 One disadvantage of photovoltaic systems is:

   a) they cannot be installed facing South

   b) there is little or no output at night

   c) they will always require roof strengthening

   d) they cannot be exposed to snow or ice.

20 The minimum recommended height of wind turbine blades from the ground is:

   a) 1–2 m

   b) 3–4 m

   c) 4–5 m

   d) 6–7 m.

UNIT 303
Understanding the practices and procedures for overseeing and organising the work environment (electrical installation)

Organising and overseeing the workplace environment is a skill expected of most qualified electricians. Many electricians are expected to be able to plan electrical installation work, from the initial contact with the client, through to supervising others during the construction stage, up to final handover to the client. For large contracts, these tasks might involve more than one person. For the majority of electrical installation work in the UK, however, one person will carry them out.

The key to success within the electrotechnical industry is the ability to plan and supervise effectively. Careful planning will minimise any surprises during the work, which in turn leads to timely completion with little or no additional cost. Effective supervision will lead to reduced wastage of time and materials. Each will lead to good customer relationships, improved company image and, very importantly, a motivated and contented workforce.

LEARNING OUTCOMES

There are six learning outcomes to this unit. The learner will understand:

1 the types of technical and functional information that are available for the installation of electrotechnical systems and equipment

2 the procedures for supplying technical and functional information to relevant people

3 the requirements for overseeing health and safety in the work environment

4 the requirements for liaising with others when organising and overseeing work activities

5 the requirements for organising and overseeing work programmes

6 the requirements for organising the provision and storage of resources that are required for work activities.

In this book, the coverage of learning outcomes and assessment criteria is not always in the sequence of the assessment checklist, which you can find on pages 246–250. It has been organised to provide a route through the content, that best links together for overall learning.

This unit will be assessed by:

■ scenario-based assignment (open book)

■ written examination.

Understand the types of technical and functional information that is available for the installation of electrotechnical systems and equipment

Understand the requirements for organising and overseeing work programmes

1.1 Specify sources of technical and functional information which apply to electrotechnical installations

1.2 Interpret technical and functional information and data

1.3 Identify and interpret technical and functional information relating to electrotechnical product or equipment

5.1 Describe how to plan:
- work allocations
- duties of operatives for whom you are responsible
- coordination with other services and personnel

5.5 Identify how to determine the estimated time required for the completion of the work required, taking into account influential factors

5.6 State the possible consequences of not
- completing work within the estimated time
- meeting the requirements of the programme of work
- using the specified materials
- installing materials and equipment as specified

5.7 Specify methods of producing and illustrating work programmes

SmartScreen Unit 303

Handout 2

Within these learning outcomes, you will look at the complete range of technical information, their sources, and how to identify and interpret this information. You will also explore the requirements for work sites that need to be considered when planning and overseeing work.

SOURCES OF TECHNICAL INFORMATION; ORGANISING AND OVERSEEING THE WORK

When planning, installing, commissioning and handing over electrical systems, a vast range of technical information is used. This information may include:

- statutory regulations
- non-statutory regulations
- codes of practice
- manufacturer's data and information
- wholesaler's/supplier's data and information
- client and installation specifications
- information from employer organisation
- contract information such as drawings, charts, graphs
- handover information.

Statutory and non-statutory regulations are covered in detail on pages 218–225. Codes of practice are covered on pages 228–229.

Manufacturer's data and information

Manufacturers are responsible for providing a vast range of technical information, including:

- product specifications
- installation instructions
- user information
- safety certification.

Product specifications

It is essential that equipment is fit for purpose. Before any equipment is purchased, the buyer must check the product specification. Many manufacturers will provide this information free of charge, either in the form of a printed catalogue or data sheet, or in the form of an electronic version such as a web page or downloadable file. In rare situations, manufacturers will only provide this information following the buyer making contact with a member of their sales team.

Product information will normally contain essential technical information. Depending on the product this may include the following:

- **Dimensions** – essential when considering the location of equipment.
- **Weight** – essential when considering possible fixing methods or support.
- **Loading** – essential for determining circuit requirements such as cable cross-sectional-area, protective devices, maximum demands and power factor.
- **Material** – what the product is made of and whether the material will react with adjacent materials.
- **Connection methods** – many items or products may require specialist methods of connecting cables and terminating the conductors. This, in turn, affects the selection of cable used to supply the equipment determines any specialist tools required.
- **Output/performance** – helps you determine whether the product will perform as required and actually do what is intended. This information may be in the form of a description or, in the case of lighting, photometric data.
- **Finish** – helps you determine whether the colour or material finish is suitable aesthetically and suitable for its intended location.
- **Associated equipment** – helps you determine whether you need to purchase any further products in order for the product to perform correctly and safely.
- **Environmental suitability** – helps you determine whether the product has a suitable IP rating or protection for the intended environment.
- **Temperature** – helps you determine whether the product will pose a fire risk due to high operating temperatures, and whether the product is affected by the location's ambient temperature.

All of the above information must be carefully scrutinised before any product is purchased, as an incorrect product could be costly to replace or modify.

ASSESSMENT GUIDANCE

Wholesaler's catalogues on the internet are an endless source of information. The information is constantly updated, unlike a brochure which is often out of date as soon as it is published.

Installation instructions

All but the simplest products will have installation instructions. These are normally packed with the product but may be available from the manufacturer before the product is purchased. The installation instructions will normally provide key detail such as suitable methods of fixing and how to connect the product.

A very important factor to note is that the manufacturer's instructions will normally supersede the requirements of BS 7671. An example of this could be that a manufacturer stipulates that a particular product should be protected by a 30 mA residual current device (RCD), yet BS 7671 may not require the particular product or circuit to be protected by an RCD. In this situation, the manufacturer's instructions should be followed over and above the requirements of BS 7671:

> 134.1.1 Good workmanship by competent persons or persons under their supervision and proper materials shall be used in the erection of the electrical installation. Electrical equipment shall be installed in accordance with the instructions provided by the manufacturer of the equipment.
>
> (IET BS 7671 Regulation 134.1.1)

Installation instructions should ideally be researched and understood during the planning stage of any electrical installation in order to determine whether any further equipment is required.

Installation instructions should ideally be kept by the installer and passed on to the client, within any operation and maintenance manual for that installation, as it may contain essential commissioning and maintenance information required later.

User information

It is essential that installers of equipment pass on any user information provided by a manufacturer, to the client. This will also form a critical part of an operation and maintenance manual. What may seem a simple-to-use item to the installer may prove complicated to the end user. An example of this is a time-switch. The installer may demonstrate the operation of this to a client or user but the user may quickly forget the verbal instructions.

User instructions or information will also contain information on any necessary spares or maintenance requirements for a particular product.

Safety certification

Any electrical product must carry a mark, label, or have associated certification supplied with it. This is to show that the product has been appropriately manufactured and tested to the product's minimum safety standards. Caution must be exercised if you are asked to install

---

ASSESSMENT GUIDANCE

Technical information should be supplied to the client for future reference.

a product that has no visible safety certification, if you are not sure of the product's origin.

Wholesaler's/Supplier's data and information

Many suppliers or wholesalers will pass on manufacturer's data, but many also produce their own paper-based and online catalogues. An electrical wholesaler will normally be able to provide verbal technical information about particular products, and will have knowledge about any additional products that may be needed when installing any specific item of equipment.

A big advantage of using wholesaler's information is the ability to compare products from a range of manufacturers. There is, however, the disadvantage that the full specification for a product may not be published.

ACTIVITY

The internet is an excellent source of up to date information. Look up the details of a 32 A builders' supply unit and copy down the specification (current rating, dimensions, etc).

Client and installation specification

Job specifications are compiled in a variety of ways depending on the size, complexity and contractual arrangement of a job. Generally the specification for a project includes the following sections.

Preliminary information

This first section includes details of the client, the contract administrator, the form of contract governing the project (eg JCT Minor Works 2011) the project and details such as the anticipated overall contract period.

In order to give some indication of all the different trades involved, a construction programme may also be referred to in the preliminaries. In large multi-trade projects the preliminaries also include the cost of site set-up and maintenance of items such as welfare and storage cabins, given as a cost per day or per week.

Scope of works

The preliminaries are normally followed by a scope of works. This section determines the extents of the specification (ie the extent of what the contractor has to quote a price for). If the contract is agreed, the scope of work will become the extent of work to be carried out.

For example, a simplified scope of works may be: The scope of works for this project is to install electrical systems within a new two-storey extension to an existing four-bedroom house. The works will include all wiring and containment systems for power, lighting and distribution. The works will not include the intruder alarm or fire detection systems.

Although this scope of works immediately eliminates the fire and intruder alarm detection, there will be interfaces between systems that need to be defined. These are generally included in the particular specification and drawings that follow this section.

Specification of works

This section describes the particular works to be carried out. It gives as much detail as possible of what the designer and client require for that particular package of works. It may consist of simple descriptions or technical requirements laid down for the contractor (tenderer) to price and eventually work to.

Materials and workmanship

Also known as the standard specification, this section is usually the same for all projects. It comprises a list of requirements in terms of workmanship and selection of material grades and quality, that applies to each job, along with any British Standards or industry codes of practice that may be applicable.

Schedules

This section includes drawing issue sheets indicating which drawings have been issued as part of the tender/contract. It also contains other schedules such as equipment schedules containing manufacturer selections, performance requirements, sizes and weights, and distribution board schedules.

Schedule of costs

This is usually the final document in the specification. The contractor (tenderer) has to complete it, specifying the cost of sections of work. Day work rates also have to be specified in case unmeasured or urgent works need to be carried out, which are not already contracted and cannot reasonably be priced or measured because of time pressures.

Information from employer organisation

During the planning and construction of an electrical installation, much information will be supplied to the site from your employer. This will include the client information and specification as detailed above (page 181), drawings as detailed below (pages 183–188) and also information relating to the administration of the organisation. This includes:

- copies of purchase orders so deliveries can be checked to ensure they are correct in quantity and type
- generic risk assessments and method statements. Many tasks that are carried out on site are repetitive and common to all construction sites. The employing organisation will have devised generic risk assessments and method statements which should

require minimal revisions to suit the particular construction site. This saves a lot of time for the supervisor on the site, meaning that risk assessments and method statements only need to be completed from scratch for tasks that are unique to that site

- further health and safety information, such as COSHH data sheets
- regular updates and reports relating to labour, such as sickness or labour being diverted to other jobs.

Contract information such as drawings, charts and graphs

Numerous sets of drawings are produced to communicate information about individual systems or collections of systems within a project. They include:

- plans/layout drawings
- schematic (block) diagrams
- wiring diagrams
- circuit diagrams.

Other types of information, in the form of graphs and charts, which assist in the planning and construction of an electrical installation are:

- work programmes
- critical path networks.

Drawings usually use British Standard symbols. However, in order to represent the wide variety of materials and equipment available, non-standard symbols may also be used. A legend or key explains them.

Plans or layout drawings are used to locate individual systems within the overall project and give an indication of the scale of the project. In addition, there may be drawings to show specific fixing, assembly and/or completion details. This is often the case where complex construction, lifting or use of a crane is required. These details may be provided in elevation, plan or both. They will then become part of the contractor's method statement when the project goes into the construction phase.

Schematic (block) diagrams can serve many purposes, but are primarily provided to show the overall functionality of a system, including interfaces and operational requirements.

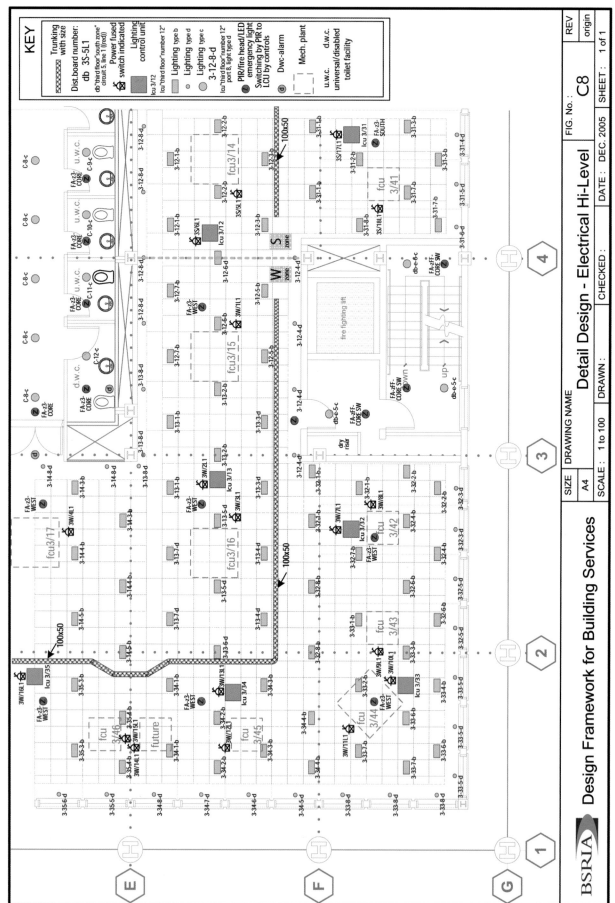

Typical layout drawing (high-level electrical pictured here)

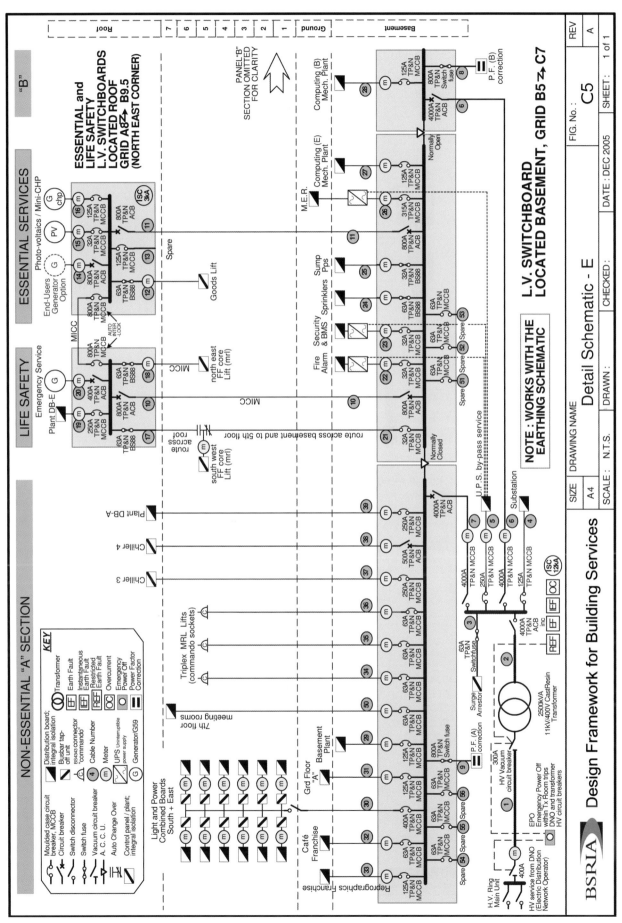

Typical schematic drawing

Wiring diagrams are generally provided to show in detail how a system or collection of systems is put together. This type of drawing shows locations, routing, the length of run and types of systems cabling. These types of diagrams are sometimes mistakenly referred to when it is actually a circuit diagram that is required.

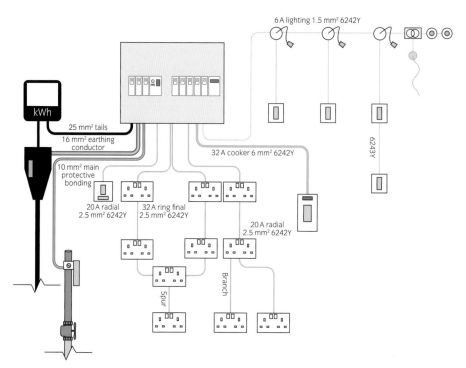

Typical wiring drawing

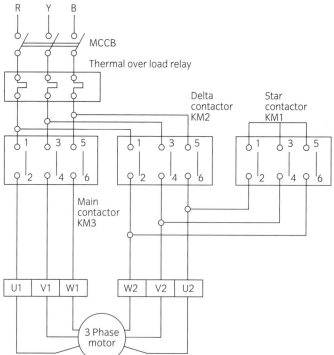

Typical circuit diagram

Circuit diagrams contain information on how circuits and systems operate. They can be provided as detailed layouts, although some information such as length of run may be omitted for clarity. In most instances, these diagrams are used for diagnostic purposes so that designers, installers and maintainers understand how their actions may influence a particular component or arrangement.

At different stages of work, the information on the drawings and diagrams is given in different levels of detail. The information becomes more detailed and accurate as the project develops.

Levels of drawing

There are five main levels of drawing:

- sketch drawing/information
- design/tender drawings
- contract drawings
- working/construction drawings
- as fitted/installed drawings.

Sketch drawings are often hand-drawn sketches to help the designer demonstrate ideas and interpretations. They are usually drawn in the early stages of a project when the scheme and designs are still fluid. They are therefore subject to change as every stakeholder is inputting to and potentially changing the scheme.

Sketch drawings are also often used at later stages of a project. Once the contractors have produced their working drawings, the designer may issue new sketch drawings for particular design details. These drawings become a contractual variation for that portion of the works and are recorded as a variation by the contract administrator.

Design/tender drawings are produced by a designer to convey the main design intent of the works. Although they are generally intended to be read in conjunction with the specification and other written documents, it is often easier to show the geographical scope of works by hatching out areas not covered on a drawing.

Design drawings are usually exchanged between designers initially to ensure coordination and then to communicate items for consideration under the Construction (Design and Management) Regulations 2007 (CDM 2007), to other designers and stakeholders, including the client.

At the design stage of a project, the designer has a duty to eradicate hazards (whether notifiable under the CDM Regulations or not) where possible. A number of hazards may be unavoidable and these need to be pointed out to contractors. Residual hazards should be identified on the drawings or schedules. The Health and Safety Executive (HSE) prefers residual hazards or hazard reduction strategies to be indicated on drawings.

The design drawings are then put together with any stakeholder comments and issued with the specifications and schedules as tender drawings. Details of the residual hazards and hazard reduction strategies on the drawings help contractors (tenderers) to understand any potential difficulties or additional requirements for completing the task safely. Allowances can then be incorporated into the pre-construction health and safety plan so that construction is carried out in a safe manner.

Following tender award, the tender drawings usually form part of the contract on which the scope and detail of work has been priced. These drawings are usually re-issued, containing any contractual variations made during the tender period. They are then referred to as **contract drawings**.

On many smaller projects the general contractor usually refers to these drawings as construction drawings (ie drawings of what the contractor tendered for and will install). However, in building services engineering, contractually there is a considerable difference between a design drawing or a briefly amended contract drawing and a construction/working drawing.

Construction/working drawings are usually produced by the building services sub-contractor. They are usually drawn at a very large scale and contain additional information not given on the previous drawings, for example on fixings, expansion measures, etc. This level of detail allows installation exactly to the drawing. However, it is important that these drawings also reflect the overall design intent.

Specialist shop drawings are a variant of construction drawings. They are normally produced so that manufacturers, suppliers and/or consultants can agree before manufacture takes place.

In building services, **as installed/fitted drawings** are normally produced by the sub-contractor/installer. However, in structural engineering and certain architectural arrangements, they are provided by the structural engineer and architect respectively. This difference can cause problems and confusion.

These drawings should represent what has been installed, but also contain information on any residual hazards and risk reduction strategies employed and instructions on how to clean, maintain or deconstruct the works if necessary. They become part of a wider set of operation and maintenance manuals.

Work programmes

A work programme is initially drawn up by the client or the client's quantity surveyor to help determine the construction period and whether the scheme is feasible and affordable in the client's timeframe.

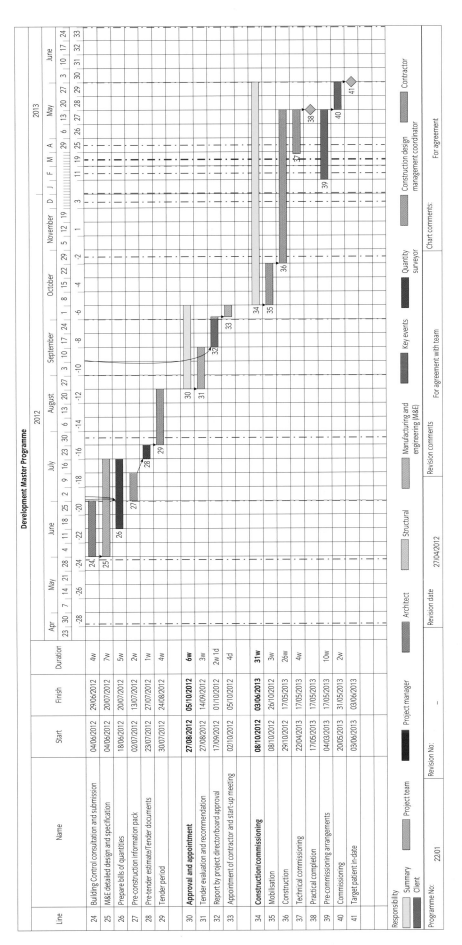

A typical client team work programme in its final stages

Once a contract has been awarded, the responsibility for the work programme lies with the contractor. On larger projects and in larger construction companies, specialist planners are dedicated to producing large construction programmes.

Specialist software collates the detailed stages of hundreds of activities, individual start and completion dates, drying and curing times, materials procurement and so on, to find the most cost-effective timeline for achieving the construction, which usually involves the shortest period on site.

Time on site has a direct impact on the contractor's costs – quite literally, time is money. So it is important to determine when the number of men on site can be reduced and the welfare, cabins and storage reduced accordingly.

Critical path

Events and activities make up the **critical path** and must be completed before a set of activities can start. For example, a building needs to be watertight before wiring can be started, which makes being watertight a critical point on the critical path.

On larger construction sites, a work scheduling technique called the 'critical path' method (CPM) or critical path analysis is adopted. The purpose of this method is to determine the shortest time in which the project can be completed, by sequencing the various construction activities. The outcome is the critical path that will lead to the minimum construction time.

Once a critical path analysis has been carried out, the project manager may decide there is too much risk in a particular critical point and ask the planner to replan activities to remove it from the critical path. This involves building contingency time around that point. This will result in a slightly longer programme, but adds an element of realism so that if delays occur around the identified point, they do not affect the critical path.

The basic critical path technique is to:

- list all the activities necessary to complete the project
- add the time that each activity will take
- determine the dependency between the activities; for example, activity C cannot start until activity B is completed.

The analysis will identify those activities that are essential – that generally take the longest time to complete and that form the critical path around which the other activities are to be organised.

Critical path

The sequence of key events and activities that determines the minimum time needed for a process such as building a construction project.

In its simplest form, the critical path below shows how the sequence of tasks effect each other.

From the diagram, it can be seen that the critical path is the longest route, which is A-D-E, as that takes nine days. Tasks B-C takes eight days and A-F takes seven days.

If task F, during the planning stage was extended to six days, perhaps due to a specific curing or drying period, the whole job would be extended by 1 day and the path A-F would become the critical path.

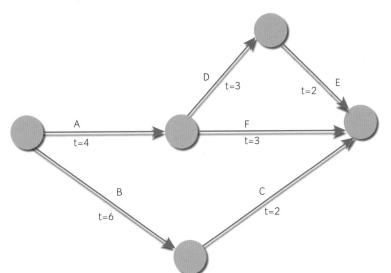

The job has 6 activities

Activity A takes 4 units of time and is not dependant on any other activity

Activity B takes 6 units of time and is not dependant on any other activity

Activity C takes 2 units of time and is dependant on completion of B

Activity D takes 3 units of time and is dependant on completion of A

Activity E takes 2 units of time and is dependant on completion of D

Activity F takes 3 units of time and is dependant on completion of A

The critical path of activity A through D and E and gives a minimum time of 9 units

Activity F is sub critical and has a float of 2 units

Path B through C is sub critical with a float of 1 unit

Critical path method

The example above states the task details in the connecting lines. Other critical path networks may show the task detail in the circles.

If a contractor or sub-contractor fails to work to the specified schedule, they may face financial penalties which would be set out as a penalty clause in the contract. Although programmes of work do allow for some potential delays, for example due to weather, one of the main benefits of using programmes of work is that the programme can be adjusted in order to minimise delays.

Handover information

Handover information needs to be delivered in order to achieve **practical completion**. As buildings have become more sophisticated, operation and maintenance documentation, for example, has become even more important. Handover information should include:

- a design philosophy and statement to say how the building has been designed and how it is to be operated, often now part of the building logbook

- Building Regulations tests and approvals, including the sign-off documents from the Building Control officer or approved inspector

- building leakage test certificates from construction

Practical completion

The point at which a construction project is virtually finished, when the last percentage of monies are paid by the client. The responsibility for insuring the construction transfers to the client and the architect or contract administrator signs the certificate of practical completion. At this point there will still be a few minor snags to sort out (eg scratches in paintwork) or insignificant items to be completed.

SmartScreen Unit 303

Worksheet 2

- the energy performance certification and registration information
- electrical test information
- lighting compliance certification (eg compliance with CIBSE LG7)
- emergency lighting test certification
- fire alarm certification
- emergency voice communication (EVC) or disabled refuge system certificate
- construction plans at a scale of 1:50 or 1:100 as appropriate
- building envelope (fabric of the building) details
- structural plans with floor loadings, any specialist support information for plant, etc
- mechanical and electrical equipment layouts at a scale of 1:50 or 1:100 as appropriate
- plant and specialist area layouts at a scale of 1:50, or even 1:20, depending on complexity of the systems
- additional information relating to the cleaning, maintenance and deconstruction of any of the systems or components and any disposal requirements at the time of writing the handover information.

It is most important that the documents are up to date and specific and relevant to the building or project being handed over.

WORK SITE REQUIREMENTS AND PROCEDURES

Before work begins on a construction site, the site needs to be ready for a variety of activities and persons. This includes:

- service provision
- ventilation provision
- waste disposal procedures
- storage facilities
- health, safety and welfare requirements
- access and egress.

Service provision

For a construction site to function, services are required which include:

- electricity supplies
- wholesome water and other water sources
- waste water systems
- fuel sources.

Electricity supplies

Nearly all construction sites will require an electricity supply for power tools and lighting, as well as for administration and welfare blocks. Where the site is a refurbishment project, the existing electricity supply may be used but careful consideration must be given to the requirements of Section 704 of BS 7671:

704.411.3.1

A PME earthing facility shall not be used for the means of earthing for an installation falling within the scope of this section unless all extraneous-conductive-parts are reliably connected to the main earthing terminal in accordance with Regulation 411.3.1.2.

NOTE: If the PME earthing facility is considered for use, see also BS 7375.

From BS 7671:2008 (2011) Requirements for Electrical Installations, IET Wiring Regulations

As most major construction sites which use the existing electricity supply cannot guarantee the reliable connection of the Main Protective Bonding Conductor during refurbishment work, the site supply may be converted into a TT earthing arrangement during the site work. This means that the earthing for the construction site will rely on an earth electrode. An assessment must also be made as to whether the existing supply is suitably rated for the intended load during construction (see the considerations for load below)

If the construction site is new or remote then a new supply must be requested, or a generator obtained. In both cases, an assessment of the demand needs to be made. This is difficult as the total load will change regularly. A good way to gauge the overall load is to assess the overall kVA (kilo volt-amperes), taking into consideration:

- the quantity and rating of site transformers used to provide 110 V supplies
- charging points for battery-operated equipment
- total temporary lighting loads
- large plant and machinery such as cranes, welders, bench saws etc.

If the site is considerable in size, a three-phase supply may be required.

If a new supply is required, an application needs to be made to the local distribution network operator (DNO). This application needs to be made well in advance of work commencing, as arranging and installing a new electricity supply can be a very lengthy process.

ACTIVITY

Look around the construction site where you are currently working. How is the site lighting arranged (type, voltage)?

KEY POINT

All generators have a rated output measured in kVA not kW. Generators can be seriously overloaded if this is not considered.

Application to install a temporary electricity supply

If you would like a temporary electricity supply and have gone through the stages outlined on previous page please complete your application here.

Site details Where you need the work done

Name: _____ Company name: _____
Address: _____
_____
Postcode: _____ Email: _____
Telephone: _____ Mobile: _____

Customer Correspondence details (if different from site address)

Name: _____ Company name: _____
Address: _____
_____
Postcode: _____ Email: _____
Telephone: _____ Mobile: _____

How would you prefer to be contacted by us? ☐ Email ☐ Phone ☐ Letter

When do you need the work completed? (This gives us an idea of your anticipated timescales) Month: _____ Year: _____

Would you like to receive your quote by email? ☐ Yes ☐ No

Property details

Is there enough parking for two of our vans? ☐ Yes ☐ No

Please make us aware of any issues restricting access to your site. For example, parking permits, double yellow lines etc.

_____

Your project

Required power: _____ kVA / kW (delete as appropriate) ☐ Single phase (up to 23kVA) ☐ Three phase (up to 70kVA)

Load Details
Please provide details of any air conditioning, heat pumps, water heaters, lifts, motors, refrigeration, welders, cranes, swimming pools, power showers, under floor heating or any other pieces of equipment that are running high loads.

Type of appliance (e.g. motor, welder, heat pump, swimming pool)	Rating of appliance	How often will the appliance be started in one hour?	Single or three phase?	Starting method (Star-Delta, Direct-on-Line, Soft Start)	Starting Current
	kW				amps
	kW				amps
	kW				amps
	kW				amps

⇨ Submit

If the above criteria don't fit your project please call us on **0845 234 0040**

Please also include:
To enable us to accurately quote your work we need a clear plan showing where you want to locate your TBS.

It would also help us if you could identify your site boundary and let us know of any hazards or obstructions at site.

Location map

Site map

Submitting your application
Save this form then mail it to:
smallservices@ukpowernetworks.co.uk

Print the form and post it to:
Small Services Department,
UK Power Networks, Metropolitan House,
Darkes Lane, Potters Bar, Hertfordshire EN6 1AG
or Fax it to: 08701 964133

The UK Power Networks' application for installation of a temporary electricity supply

ASSESSMENT GUIDANCE

This form and other information is available on the internet.

ACTIVITY

Using the internet, find out roughly how long it takes to obtain a temporary electricity supply from initial application to installation.

Water bowser

A transportable water tank or tanker.

Wholesome water and other water sources

Water is essential on site. Wholesome water is drinking water and is necessary for the welfare of those on site. Other reasons for requiring a water source are:

- mixing cement or other materials
- washing or showering
- cleaning of equipment, roads or structures
- dealing with hazards, dust or chemicals.

In remote locations, **water bowsers** could be used to provide a water supply if mains water is not available. Water should not be drawn from rivers or lakes unless specific permission is granted from the Environment Agency.

Waste water systems

Waste water needs to be considered along with water supplies as mentioned above. Waste water systems need to be looked at in two ways. Firstly, any surface water from rainfall needs to be diverted into local watercourses or **soakaways**. Secondly, **blackwater** needs to be discharged into either main sewers or septic tanks.

Fuel sources

Larger construction sites will need to consider storage of fuels, such as diesel, for plant, vehicles and generators.

Ventilation provision

Ventilation needs to be considered in situations such as confined locations, tunnels, underground works or, more generally, in large buildings. Ventilation should also be considered for specific tasks which involve a large amount of dust, where the area is to be sealed off.

Ventilation may also be required for particular items of equipment that are vulnerable to dust contamination.

The most common method of ventilating areas is by a forced clean air system, where clean air is forced by fans through ducting into the area and, in many cases, extraction fans then remove stale air from the location.

Other locations and equipment may adopt a system where air is forced into an area or enclosure creating a high pressure. This high pressure minimises contamination by keeping out dust and contamination from lower pressure areas.

Waste disposal procedures

Careful consideration needs to be given to areas set aside for waste storage and collection. As waste is normally collected by large vehicles, easy access is one of the most important factors that effects the location of waste storage. Methods of separating waste for recycling, or hazardous waste are covered in detail in Units 301 and 302.

Construction sites for high-rise buildings require methods of moving waste from upper floor levels to waste collection points at ground level. This is normally done by rubbish chutes, which are flexible ducts that allow waste to fall safely, directly into a collection point.

Soakaway

A system where rainwater or surface water is collected, then slowly discharged into the ground. The soakaway system would normally be below topsoil level to avoid waterlogging.

Blackwater

Water that is contaminated by sewage or chemicals. Blackwater generally comes from kitchen sinks and toilets.

ACTIVITY

How is cardboard packaging disposed of on site?

Storage facilities

Good storage facilities must be provided for plant, machinery and materials in order to minimise:

- theft
- damage
- congestion.

A secure lockable area, such as a storage container, is a good method of reducing theft from site as well as reducing the likelihood of damage from weather or general site activities.

If a good secure area is not set aside for storage, materials left around a site can become a hazard by potentially blocking escape routes or becoming obstacles to others trying to carry out their tasks. As well as being a safety hazard, poor storage also becomes inefficient as it is more likely that materials will become damaged or lost, or time is wasted moving the materials out of the way.

When locating storage facilities, consideration should be given to:

- good access for deliveries
- good access for taking materials to site
- avoiding locations where work is to be undertaken
- security
- avoiding blocking escape routes or routes for emergency services.

Health, safety and welfare requirements

All construction sites, regardless of their size, must provide basic amenities such as:

- toilet facilities
- washing facilities (hot and cold water)
- first aid facilities
- changing rooms or lockers.

On larger construction sites, rest areas which provide shelter from the weather should also be provided. Showers must be provided on sites where tasks involve working with chemicals or hazardous materials.

When considering welfare facilities, the Health and Safety Executive has produced an order of preference for the provision of welfare facilities, dependent on the nature of the construction, as shown in the following table.

ASSESSMENT GUIDANCE

Basic first aid facilities are essential on any site. They should not contain tablets as these can only be dispensed by a qualified medical person.

Type of installation	Additional notes
1a Fixed installation: connected to mains drainage and water.	Order of preference:
1b Portable water flushing units with water bowser supplies and waste storage tanks.	■ on site ■ at a base location ■ at a satellite compound. NB This may include the pre-arranged use of private facilities. Permissions, preferably in writing, should be obtained from the proprietor in advance of the work starting. The use of public toilets is acceptable only where it is impractical to provide or make available other facilities.
2 Portable installation on site.	Consisting of chemical toilet(s), washing facilities and sufficient tables and seating.
3 Suitably designed vehicle.	Consisting of chemical toilet(s), washing facilities and sufficient tables and seating.
4 Facilities which are conveniently accessible to the work site (includes public toilets).	Use of public toilets is acceptable only where it is impractical to provide or make available other facilities.
5 Portable installation near site.	Incorporating a chemical toilet, washing facilities and sufficient tables and seating.

Taken from Construction information sheet no. 59, Health and Safety Executive (HSE)

Access and egress

Suitable provision for safe and controlled access and **egress** is required for all construction sites. It is essential to be able to monitor who is on site, whether they be working at or visiting the site. This ensures that:

■ all people can be accounted for in an emergency

■ information or instructions, resulting from ever-changing conditions or dangers, can be safely and speedily distributed

■ the security of the site is closely controlled

■ the safety of all on-site personnel, or members of the public who may accidently stray onto the site, is closely controlled.

On most sites, all personnel and visitors should report in and sign an attendance record book. Construction site safety boards are usually located at the entrance points providing instructions on the basic minimal safety requirements.

All sites must also provide suitable evacuation points which allow all persons to leave the site quickly and safely and go to a designated assembly point, should any emergency situations occur.

Many large construction sites will have security personnel who monitor all who enter or leave the site.

ACTIVITY

Check the HSE website to see what guidance is available for providing welfare facilities on site.

Egress

Leaving or exiting the site.

ASSESSMENT GUIDANCE

'Egress' may not be a word you are familiar with. It is the opposite of 'access'. Provision for safe egress means providing a safe and easy exit route.

Assessment criteria

1.5 Identify equipment and systems that are compatible to site operations and requirements

ACTIVITY

A single-phase 110 V system has a voltage to earth of 55 V. What is the voltage to earth on a three-phase 110 V system?

COMPATIBLE EQUIPMENT AND SYSTEMS

It is not uncommon for the installation electrical contractor to be responsible for on-site temporary electrical supplies within a construction site. Site temporary supplies may be a simple single-phase consumer unit with RCBO protection supplying a number of socket-outlets on a single board, known as a 'builder's board'. From these socket-outlets, site operatives will plug in their 2 kVA 110 V centre-tapped earth (CTE) transformers in order to supply their power tools or lighting. This arrangement is suitable for small refurbishment type projects or the construction of small buildings.

On major construction sites, much larger site temporary systems are usually installed which provide 110 V CTE supplies throughout the construction site as well as suitable lighting. This involves taking 230 V supplies through the construction site to supply large multi-way 110 V CTE transformers. The table below shows the type of electrical supplies usually required on a construction site together with compatible wiring systems.

Nature of supply	Type of suitable cable	Reasons for selection	Notes
Supplies to remote 110 V CTE transformers	Steel-wire armour cable	Very good mechanical protection, as supplies will be 230 V to earth	Must be protected by a 30 mA RCD
Supplies to fixed temporary lighting	▪ Steel-wire armour cable ▪ Hituf cable	Very good mechanical protection	Must be RCD protected or 110 V CTE
Supplies to site administration or welfare blocks	Steel-wire armour cable	Very good mechanical protection	Site administration blocks are outside the requirements of section 704 of BS 7671, but the supply cables *are* covered by section 704 if they are within the designated site boundary
Supplies to power tools or portable lighting from 110 V CTE transformers	Arctic flexible cable	▪ Flexible ▪ Good protection against oils and some chemicals ▪ Good in a range of temperatures	Colour-coded outer sheath to denote voltage ▪ purple < 50 V ▪ yellow 110 V ▪ blue 230 V ▪ red 400 V

Socket outlets for power tools used on construction sites are fitted with BS EN 60309 plug and socket outlets, which are also colour coded to indicate voltage.

As well as the colour coding, plugs and sockets have notches and bumps called keys which limit their connection. So, it is not possible to plug a piece of 110 V equipment into either a 230 V or 400 V supply. If a BS EN 60309 plug has been cut off and replaced with a standard BS 1363 13 A three-pin plug, you should not use the equipment and should bring it to the attention of your supervisor.

Section 704 in BS 7671 Requirements for Electrical Installations (the IET Wiring Regulations) provides further information regarding the installation requirements for temporary power supplies and installations to be used by site workers during the construction stage.

ASSESSMENT GUIDANCE

13 A plugs and sockets have rectangular pins which will not fit into round pin accessories. The key ways on BSEN 60309 outlets have positions indicated by clock positions – 6 o'clock, for instance, being at the bottom.

SmartScreen Unit 303
Handout 11

Assessment criteria

2.1 State the limits of your responsibility for supplying technical and functional information to others

5.4 Identify, within the scope of the work programme and operations, your responsibilities

KEY POINT

Where an instruction to stop or start a particular item of work is received from anyone other than the CA, the contractor accepts that instruction at his own risk. It is therefore good working practice to ensure that any immediate instruction is followed up with a formal instruction, otherwise the contractor is perfectly entitled (except in terms of safety) to carry on regardless.

Contract administrator

The person named in the contract with the contractual power to change matters or items of work that will cause a contract variation.

Supplying clear and precise technical information is essential during construction works as missed or misunderstood instructions could lead to danger, costly mistakes and poor reputation. In this learning outcome you will be looking at the roles and responsibilities within the construction industry and the methods used to communicate in the best way with each person.

THE KEY ROLES OF THE SITE MANAGEMENT TEAM

To understand how to communicate efficiently within the construction industry, an understanding of who to communicate with is essential.

It is important for everyone to understand the different roles and responsibilities in a site management team. Some of the roles and titles differ slightly depending on the individual organisation or the size of project. The following are examples of the main positions.

Architect

The architect has traditionally been the individual associated with overall responsibility for a project. He or she is normally the one person accredited with the construction of a particular building, for example, St Paul's Cathedral in London, built by Sir Christopher Wren.

The architect is usually employed by the client. In many traditional construction contracts, the architect takes the role of design team leader and is named as the person with authority to vary the works by use of an architect's instruction (AI). Modern non-traditional contracts allow others, such as the client, project managers, etc, to take the role of **contract administrator** (CA), who is also allowed to vary a contract by instruction.

The title 'architect' is a legally protected title, which is enforced by the Royal Institute of British Architects (RIBA). Other persons carrying out architectural design works are usually architectural technicians or architectural design consultants. These individuals have not normally completed the seven years of formal training required by the RIBA and are not officially entitled to use the title 'architect'.

The Swiss Re building, known as the Gherkin and designed by internationally renowned architect Sir Norman Foster

The architect is usually responsible for all elements of architectural or building design including the specification of building form, structure, fire strategy, orientation, and interior and exterior design to meet client requirements, planning conditions and building regulation requirements. Generally the architect does not carry out any special mechanical or electrical design works but is responsible for the final coordination of the building services fixtures in terms of reflected ceiling plans (plans of the ceiling indicating, for ease of coordination, lights, grilles and architectural features).

In specialist projects or where the task requirements are challenging, the client will acquire specialist design from the other members of the design team including the structural engineer, landscape architect, fire engineer and specialist planning consultants. In these instances the architect is normally the design team leader.

ASSESSMENT GUIDANCE

If you are working on a large site there will probably be a large board at the entrance which details who the architect is and the other companies involved in the project.

ACTIVITY

Most organisations have structures. Think of an organisation you know, such as a sports club. Write down the organisational structure – such as the roles of the team manager, the players and coaches – to show their responsibilities and to whom they are responsible.

Project manager

A project manager (PM) is usually a building professional in private practice, with specialist project management qualifications and experience. Many PMs take a course such as PRINCE2, which is a qualification recognised by government departments for the project management of public projects. The PM's role is to ensure that the project is delivered on time, on budget and to all the correct standards.

The client normally appoints the PM to run projects and manage the works. The PM acts as the project leader and in certain contracts as the contract administrator, responsible for the overall contract. The PM issues contract instructions and generally deals with the contractor and other professionals working for the client. The PM normally chairs the construction project meeting between the client team and the contractor's construction team, reporting directly to the client or funding body.

On larger projects there may also be a PM on the contractor's construction team. This person is normally site-based and has ultimate responsibility for the success or failure of the contractor's construction project. In order to ensure delivery, the contractor's PM is the leader of a multi-disciplinary site management team. In this role, the PM normally meets with the client on matters of contract, programme and cost.

The contractor's PM is usually focused on delivery of the project on time. In order to ensure a profitable outcome, the contractor's PM usually drives the contractor's team to ensure that the critical points on the construction programme are met. Often this involves putting pressure on contractors and sub-contractors to make up lost time or on the client for information to ensure **work packages** are not delayed.

Work package

A 'collection' of work associated with one product on a project, eg all processes associated with the design, manufacture and installation of the windows or of the sprinkler installation.

Clerk of works

The clerk of works (CoW) is another representative of the client. The CoW is responsible for ensuring that work is carried out on schedule and that the quality of the works meets both the specification and industry standards, including British Standards (BS) and European Normative standards (ENs).

The CoW keeps a record of weather conditions and the labour resources on site to ensure that the work is carried out on schedule. Another responsibility is to ensure that adequate time and resource has been applied to particular tasks and that any short cuts have not impacted on quality, commissioning or warranties. The CoW also issues site correction notifications to the contractor to indicate areas of work that do not meet the contract.

ASSESSMENT GUIDANCE

You may be asked to list conditions that may result in delays of a contract. Think of ways that the weather may cause delays. Be specific on what the delay may be. Such as: high winds – stop the roof being completed.

The CoW is not normally given the power by the client or by the terms of the contract to vary or extend the works and therefore does not usually have any contractual connection to the trades on site. However, the CoW's role includes liaison and discussion with the different trade contractors with respect to the quality of works and the number of workers. The CoW puts this information into reports to the client or the client's agent (architect or project manager) and the contractor.

Construction manager

The construction manager is part of the contractor's on-site team and reports directly to the contractor's PM.

The construction manager is responsible for ensuring that the construction element of the project runs to time and to the relevant quality standards, to ensure minimal reworking. The construction manager is responsible for all site activities, including inductions and site safety, and ensuring that contractor and sub-contractor **toolbox talks** take place. The sub-contractor has the most dealings with the construction manager as he/she spends a considerable amount of time out on the site.

The construction manager has normally been trained in general construction either from a surveying, trade or construction background.

Toolbox talk

A method of communicating safety issues (as a supplement to an induction) or of giving continuous training on particular techniques or methods.

Quantity surveyor

The role of quantity surveyor (QS) can exist in the client team and also in the construction team. Both roles fulfill very similar responsibilities but from different perspectives.

The client's QS is normally a professional from a private practice of chartered quantity surveyors, which are generally regulated by the Royal Institute of Chartered Surveyors (RICS). The QS's role is to put together estimates by measuring and **quantifying** the designs from the client's design team prior to **tender**. The QS advises the client on budget allocation, selecting the method of procurement and contract documentation. In certain types of contract they are also responsible for producing a **bill of quantity**. Once a project has been tendered, the QS reviews the priced tenders of the bidding contractors, checking the commercial terms and conditions and financial accuracy of each tender to ensure that the contractor has submitted an appropriate price.

Quantify

Estimate guideline costs.

Tender

An offer to carry out work, supply goods or buy land, shares or any form of asset for a fixed price, usually in competition with others.

Bill of quantity

A contract document comprising a list of materials required for the works and their estimated quantities.

The contractor's QS is normally employed within the contractor's business. Their role is to assess the tender documentation, bills of quantity, etc and check the costs that the estimating department has provided to ensure that the bid is competitive enough but still gives the company adequate profit.

The QS also checks the commercial terms of the contract. They review the contract, referring the documents to company lawyers if the terms are non-standard or appear too difficult to accept outright. This can result in negotiations over contract clauses between the contractor and the client, which are normally settled with a compromise that is acceptable to both parties.

ACTIVITY

Describe how a quantity surveyor may be involved with electrical work you may be doing.

Consulting engineers

Engineers are highly qualified, specialist personnel required on complex projects. There are many different types of consulting engineers, who work on the structure or specific services.

Structural engineer

Structural engineers are professionally qualified (chartered) in the field of structural engineering and usually part of the client's design team. The structural engineer is responsible for the structural stability of structures and ground works within the contract.

Very early in the initial design stage, the structural engineer determines the substructure of the building (ie the type of foundations and support systems). This involves the designer interpreting ground conditions to determine whether traditional foundations or piles are to be used to support the structure in the ground.

The structural engineer also designs the building support system, depending on the type and use of the building, as well as the requirements of the client. This work involves sizing and locating structural beams and determining the strength of walls and floors. On projects with no civil engineer, the structural engineer will also design features such as car parks, roads, drainage, water run-offs and environment protection measures.

Building services engineer

The building services engineer is also a professionally qualified member of the client's design team. Building services engineers are often **incorporated engineers**, although some are **chartered engineers**.

Building substructure works

Incorporated engineer

A specialist (also called an engineering technologist) who implements existing technology within a particular field of engineering, entitled to use the title IEng.

Chartered engineer

An engineer with professional competencies through training and experience, also registered with the Engineering Council, which is the British regulatory body for engineers. The title chartered engineer (C Eng) is protected by civil law.

First fix electrical installation

The building services engineer is responsible for the mechanical and electrical services in a project, often referred to as MEP. The mechanical services include heating, ventilation and air-conditioning systems. The electrical services include lighting, power and distribution, as well as telecommunications, fire alarms, alarm and communication systems, intercoms and disabled refuge communication systems. The 'P' in MEP stands for plumbing and public health engineering, including hot and cold pipework, above ground drainage and, in certain contracts, private underground systems prior to input into the public sewer. The building services engineer designs any rainwater systems that run internally, whereas the architect designs the external rainwater systems (pipes on the exterior from gutter to ground level drain).

Building surveyor

Chartered building surveyors are building professionals regulated by the Royal Institute of Chartered Surveyors (RICS) and often employed on the client's team.

They are usually responsible for assessing existing buildings or renovation schemes, including valuation of property for lenders, scheduling of building defects, refurbishments and undertaking different types of building surveys.

Land surveyor

Chartered land surveyors are also regulated by the Royal Institute of Chartered Surveyors (RICS). Their work is generally for the client and they would normally be part of the client's design team.

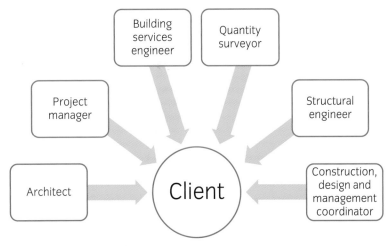

Clients' contractual relationships

They deal with the measurement of land, which may involve setting out structures on the land in accordance with the architect's drawings.

ACTIVITY

Review the websites of different types of construction-related consulting engineers. Look at the Association for Consultancy and Engineering (ACE) website. Follow the links to visit the websites of various consulting engineers. Identify the different types of consultancy that operate on a local and national level. Look at how they differentiate themselves from the competition.

They often detail the areas surveyed, so an architect can make an accurate design for a structure. They are also employed to determine the boundaries to properties, give valuations, etc.

Estimator

The estimator is part of the contractor's team and responsible for preparing the estimated price for the work in the tender package prepared by the client's quantity surveyor. This estimate has to be carefully prepared to ensure that the best prices are obtained for materials and labour from sub-contractors. The estimator is usually a former tradesman, who has moved into the estimating team.

Buyer

As the name suggests, once a contract has been won, the buyer is responsible for purchasing all the materials from suppliers, within the costs stated in the tender by the estimator. A key factor in effective tendering, which offers the most economical prices the contractor can afford to offer, is the ability to purchase or obtain the materials at the correct price, and under suitable terms and conditions, after the contract has been awarded. The buyer also has to ensure that the materials are delivered on time.

Contracts manager

The contracts manager is often a company director or reports directly to a director and works on the contractor team.

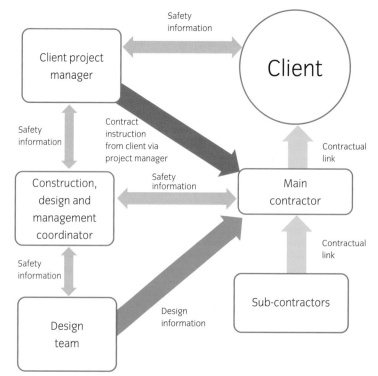

Contractor and client team relationships

Contracts managers are normally responsible for several construction projects so they are not normally part of the on-site team. As a regular visitor to site, their role is to liaise with the contractor's on-site management team, including the project manager and construction manager.

KEY ROLES OF INDIVIDUALS REPORTING TO THE SITE MANAGEMENT TEAM

Sub-contractors

Each sub-contractor is responsible for their package of work or construction discipline, such as substructure/groundwork, joinery, brickwork, window installation, electrical, plumbing, plastering or decorating. They have to work safely, to the main contractor's construction programme, producing work that meets the specification, in a timely manner and accommodates the contractor and all the other sub-contractors working on site.

In the majority of construction contracts, the sub-contractor is a 'domestic sub-contractor', who only works for a main contractor. Therefore the sub-contractor can only be instructed by the contractor as he/she has no contractual link with the client or the client's team.

> **ASSESSMENT GUIDANCE**
>
> You should understand the role of the sub-contractor in the overall building construction set up.

Site supervisor

The site supervisor is often known as the general foreman. On smaller projects, the site supervisor is the main point of contact for all trades. On larger projects, this role is generally carried out by the construction manager, to whom the site supervisor reports. In some organisations site supervisors are known as assistant construction managers.

Trade supervisor

The trade supervisor is normally the foreman or lead person for each sub-contractor. This person usually controls the individual trade operatives on site, within the general rules and requirements of the main contractor/site. They will usually be an experienced tradesman with many years of site experience, able to interpret the requirements of their own off-site management team and integrate those requirements into the wider on-site requirements to ensure smooth running between the different trade contractors.

Trades

Each of the different trades and disciplines needs to liaise with the on-site management team and others. The following are examples of different construction trades.

- **Bricklayer** – builds brick and blockwork walls with different finishes.

- **Joiner** – carries out various tasks, including constructing stud walls (usually internal timber walls) and installing doors, frames, skirting boards, cupboards and other fitted fixtures.

- **Plasterer** – applies wet plaster and finishes or provides **dry-lined walls** with specialist taped joints or minimal skimmed finish, using a small amount of finishing plaster.

- **Tiler** – a finishing trade responsible for tiled finishes.

- **Electrician** – the trade contractor responsible for the installation of all electrical wiring including any wiring containment systems or specialist systems such as door entry security or fire alarms. On smaller projects, the electrician installs all specialist services or may employ a specialist installer for systems such as the intruder alarm, to meet any contractual or specification requirements. On larger projects, the electrical contractor is usually responsible for the low voltage wiring and for providing wiring containment systems for the specialist installers of, for example, fire and security systems. These specialists are normally sub-contractors of the electrical contractor, who is responsible for managing them as well as his own staff.

- **Heating and ventilation (H&V) fitter/engineer** – carries out plumbing and heating installation works, normally in steel. (A plumber normally carries out installations in copper.)

- **Air-conditioning and refrigeration engineers** – closely aligned to the electrical trades, although carrying out some pipework installation. Refrigeration engineers have to be aware of refrigeration legislation as certain gases are banned in the UK.

- **Gas fitter** – specialist qualified to work on gas and combustion equipment installations such as boilers. Gas fitters are registered with Gas Safe, which is a competent person scheme (CPS) that ensures registered tradesmen are qualified and competent to work on gas systems and boilers, within the legal requirements.

- **Decorator** – finishing trade responsible for decorated finishes to all surfaces, including woodwork.

- **Ground worker** – manual worker associated with levelling ground, excavations for substructure and foundation works, or providing trenches for pipes or incoming services. Ground workers are usually supported by a wide range of earth-moving equipment and excavation tools from a simple pick and shovel to complex hydraulic machines such as the standard backhoe (digger).

Dry-lined wall

An inner wall that is finished with plasterboard rather than a traditional wet plaster mix.

ASSESSMENT GUIDANCE

You should be able to describe the work that each trade carries out.

ACTIVITY

Look at the list of different trades on this page and the previous one. List the order in which they may be required to start on a new house building site.

KEY ROLES OF SITE VISITORS

Building Control inspector

Every construction site is required to meet the requirements of the Building Act 1984 (as amended). The policing of this is the responsibility of the Building Control department of the relevant local authority. The planning consent may also specify additional requirements that have to be met, such as provisions for fire prevention or noise reduction.

The Building Control inspector/officer (BCO) has the power to approve or deny approval of works, construction details and layouts that affect the building, in line with the Building Regulations. Normally the BCO will refer to a particular approved code of practice (ACoP) relating to the technical area concerned. Each area of specialism is also referred to as the specific part in the Building Regulations (with the corresponding approved documents) that relates to it. For example, Approved Document P (also known as Part P) relates to the electrical installation.

An increasing number of specialist private companies now offer the services of an approved inspector. When appointed, this person has the same power as the Building Control officer.

The BCO or approved inspector normally comes from a construction background, as they deal mostly with building works. However, other areas of building regulation are classified as non-construction, as with Part P, which relates to the electrical installation. If the BCO does not have the relevant expertise, they must rely on the expertise of an approved inspector in that area.

Water inspector

Every water supplier has a responsibility to supply **wholesome** water and to ensure that numerous checks are made on the water supply. No matter how many checks are made on the public supply system, incorrectly designed and installed private systems and connections put the public water supply at risk.

Since the implementation of the Water Supply (Water Fittings) Regulations 1999, water regulations inspectors employed by the water companies have statutory powers to impose improvement orders on the users of water supplies, who can face withdrawal of the public water supply if they do not comply. Water company operations are monitored by the Drinking Water Inspectorate (DWI) water inspectors, whose duties include ensuring that water continues to be wholesome.

ASSESSMENT GUIDANCE

All site visitors should be both signed in and signed out when leaving.

ACTIVITY

Which parts of the Building Regulations apply to electrical installations?

KEY POINT

The Building Control officer does not usually have electrical knowledge or experience and must therefore rely on the electrician to provide accurate and reliable information. Competent Persons Schemes (CPSs) address this shortfall by assessing each contractor's competence in advance of registering jobs with Building Control.

Wholesome

Term used in law to indicate water that is suitable for drinking.

Compliance with health and safety law

Duty holder

The person in control of the danger is the duty holder. This person must be competent by formal training and experience and with sufficient knowledge to avoid danger. The level of competence will differ for different items of work.

HSE inspector

A health and safety inspector is an officer of the Health and Safety Executive (HSE), a government department. They have the right to enter any workplace without giving notice. They usually do make an appointment to visit, unless the element of surprise is thought necessary.

On a routine visit, an inspector would expect to look at the workplace, work activities and methods and procedures used in the management of health and safety, as well as checking compliance with health and safety law specific to that workplace. The inspector is empowered to talk to employees and their representatives and to take photographs and samples. An inspector may also call for a specific purpose, for example to follow up a complaint or incident.

If there are breaches in compliance, depending on the level of severity, the inspector can carry out the following courses of action.

- **Informal warning** – Where the breach is minor, the inspector can explain best practice and legal requirements to the **duty holder** (usually the employer) on site, following up with written advice.

- **Improvement notice** – Where the breach is more serious, this notice will say what has to be done, why and when the remedial action has to be completed by. The inspector can take further legal action if the notice is not complied with within the time period specified. The duty holder has the right of appeal to an industrial tribunal.

- **Prohibition notice** – Where there is a risk of serious injury, the inspector is empowered to stop the activity immediately or after a determined period. Once this notice is applied, the activity cannot be resumed until remedial action has taken place. The duty holder has the right of appeal to an industrial tribunal.

- **Prosecution** – In certain circumstances, the inspector may consider prosecution necessary in order to punish offenders and deter other potential offenders. In some cases unlimited fines and imprisonment may be imposed by the higher courts.

Electrical services inspector

The title 'electrical services inspector' is rarely used in the UK as the majority of installations are self-certified by the installing contractor. The inspector is normally Level 3 qualified, with specific qualifications in the inspection and testing of electrical installations.

Building services engineer

This is normally a professionally qualified individual who may have been a tradesman prior to taking advanced qualifications and training. This individual will have qualifications to carry out, for example, estimating, contract management, design or the role of specialist **authorising engineer**. Most building services engineers have undergone a training programme that satisfies the requirements of the Engineering Council, the regulatory body for engineers and technicians in the UK.

CUSTOMER RELATIONS METHODS AND PROCEDURES

A good relationship with a customer or client is very important if a business is to succeed. Within the building services industry, your customer may be a homeowner, a premises manager or another contractor who is employing your services as a sub-contractor.

Providing a professional image goes a long way to earning the trust of a client. Many organisations insist that their staff wear a uniform, which is a good way to promote this. However, the best way to present a professional image and promote relationships is to communicate clearly, efficiently and courteously. If a customer asks a question, be sure to answer in a clear and courteous manner. Above all else, the best way to earn trust is to be honest with any customer or client.

PROVIDING TECHNICAL AND FUNCTIONAL INFORMATION

Communication in the workplace is essential for the safety and well-being of staff and the profitability of a job. Various methods of oral and written communication are used in different circumstances.

Oral communication is the more immediate. For example, it can involve a word from a colleague pointing out a potential improvement, or a verbal request on site to stop work on the grounds of health and safety. It may not always be face to face but could be made via site radio, site tannoy system or telephone (for example, the client asking for a particular piece of work to be started immediately).

Oral communication is less formal than written communication. Although it can be legally binding, it is difficult to prove and can easily become confused. Therefore, most important oral communications are followed up in writing, usually referring to the date and time of the original communication.

Communication in the workplace

SmartScreen Unit 303

Worksheet 11

Written communications come in many forms, including emails, facsimiles (faxes) and letters. Fax messages were once the fastest form of written communication, being transmitted over a telephone line. This method of communication is sometimes still preferred by institutions such as banks as the contents of a letter can be seen a few days ahead of the posted version.

Most companies now communicate by email, whether on a formal or informal basis. This is a highly effective and high quality method of communicating messages, with pictures and drawings. It allows letters to be scanned, drawings to be issued and instructions to be followed up immediately. However, the ease with which an email can be sent to the recipient, and copied to many others, even within the same company, can lead to information overload and can eventually have a detrimental effect on relationships.

Letters are considered the most formal method of communication, due to the effort involved in typing, printing and posting them. They are used in the most formal situations, such as offering an appointment, terminating a contract or following up a less formal email.

Methods of communication for people with different needs

Methods of communication for people with different needs should be considered carefully and adjusted, as necessary, whilst avoiding creating awkwardness for any of the parties.

Communicating orally with most people with physical disabilities should be no different to communicating with able-bodied people. Special provision might need to be made for the visually impaired or hard of hearing. The key is to be aware and make adjustments without fuss. For example, this might involve making sure that the speaker's lips are clearly visible for lip readers to follow or adapting communication that involves visual effects or displays, to meets the needs of the visually impaired. The able-bodied person also needs to be aware that any awkwardness they communicate can become an issue.

A different type of understanding may be required when communicating with persons with learning difficulties. It is important to keep communication at a level that will be understood by the individual. If there is any doubt, it might be useful to communicate in conjunction with or through a relative or advocate, who can give feedback on whether the communication was appropriate.

Communication involving different dialects or accents is sometimes difficult. However, polite requests to repeat what has been said will usually result in the speaker using more standard English, giving the recipient a chance to become more familiar with the dialect or accent.

When dealing with a person using English as a second language, it is important to be polite and patient, and to check understanding where necessary. Use of dialect and colloquial terminology should be avoided.

Actions to take to deal with conflicts

Conflicts between clients and site operatives should be unusual. However, they are not impossible. In these situations, the company needs to ensure that the client is not offended. More importantly, it needs to ensure that the operative has not had to deal with a client who is angry with the organisation rather than them personally. In employment law, this could be deemed as harassment due to the client's unreasonable response to a breakdown in communication from the company.

Where a dispute is clearly between individuals, each party must be heard in order to find common ground or grounds for making amends. Where the dispute cannot be resolved, it may be necessary to remove the operative from the project and away from any further conflict, without blame (assuming no blame exists). In these situations, it is important to confirm politely to the client that the operative has been moved with no blame attached, in order for both parties to move on.

Where conflict exists between co-workers, there is usually a prescribed route to resolution within the company's disciplinary and grievance procedures. This normally involves an opportunity to resolve matters informally, where the two parties air their views and problems, and hopefully reach a resolution. Where this is unsuccessful, the company must get involved formally. It needs to avoid the charge of constructive dismissal if one of the parties resigns because of the dispute.

The company will try to resolve the issue. However, it might be appropriate to put both parties on a first-stage warning to stop the activity that is causing the problem. Disciplinary action will need to be taken if offensive behaviour continues, in line with the bullying and harassment policy.

This procedure would also apply in cases of conflict between a supervisor and an operative. An exception would be if the disagreement resulted from a lawful and reasonable request by the supervisor for the operative to carry out tasks. In such a case, the company would be entitled to start disciplinary procedures against the operative. Failure to carry out instructions would be misconduct, leading to a formal written warning. Repeated misconduct would lead to dismissal.

> **ASSESSMENT GUIDANCE**
>
> It is always better to resolve a problem at an early stage, as ultimately it will have to be resolved.

The effects of poor communication

Poor communication between operatives is likely to create a difficult working environment and lead to losses in productivity. Lack of communication can also impact on safety and will stifle ideas for improvement in working methods and practices.

Management has a responsibility to communicate certain matters such as safety information. Effective communication is two-way, so when communication is poor, messages are potentially misunderstood. Where there is misunderstanding, there is a risk of mistrust and where there is mistrust, there is a risk of conflict. Therefore poor communication from management can lead to mistrust and conflict, which can impact on productivity and safety.

Customer relationships are an essential part of business development and customer retention. It is therefore essential to have effective, meaningful communication with customers. Good communication between the company and customer ensures that invoices are paid on time and that both parties are prepared to do business together in future.

> **ASSESSMENT GUIDANCE**
>
> If you do not understand what has been said to you there is no shame in asking the person to repeat it. It would be worse if you carried out the instruction incorrectly.

Assessment criteria

2.6 Describe methods for checking that relevant persons have an adequate understanding of the technical and non-technical information provided, including appropriate health and safety information

WORKING POLICIES AND PROCEDURES

Companies set out working policies and procedures so that all workers can expect to be treated fairly and with respect, know how they can put right a problem should one arise and understand what measures or sanctions will be imposed should a breach of policies and procedures occur.

Working policies and procedures set out a company's expectations on matters such as behaviour, timekeeping and dress, and become part of the terms and conditions of employment. They are often provided in a staff handbook, although such a handbook is not required by law. An employee accepts the terms and conditions of employment when they sign a contract of employment or attend work without challenging any of the terms of the contract.

Contract of employment

A contract of employment is a legally binding agreement between employer and employee. It can be a verbal or written contract, including various terms and conditions. Any changes to the terms of employment will probably not be notified in a new contract of employment, but in the staff handbook, in a letter or on the company notice board.

If the contract is verbal, the employer must give a statement of the main terms and conditions within two months of the start of employment, under the provisions of the Employment Rights Act 1996. The statement would include:

- holidays entitlements
- scale of pay
- basic working hours
- pensions and pension schemes
- notice of rights.

Health and safety policy

All companies with five or more employees are required by law to set out their health and safety policy in writing. Related issues detailed in this policy, or separately, include:

- bribery
- bullying and harassment
- disciplinary and grievance procedures
- equality and diversity
- maternity, paternity and adoption
- pay
- redundancy.

It is a legal requirement for employers to set out disciplinary rules and grievance procedures in writing and make sure that every employee knows about them and the available courses of action if they are unhappy with a disciplinary decision. Although related, the bullying and harassment policy is often a separate document because of the many issues and sensitivities that have to be considered.

> **ASSESSMENT GUIDANCE**
>
> You should be able to explain a situation where redundancy may be payable.

Limits on authority and responsibilities

Within company policy there is likely to be a statement of the limit of authority for different individuals and groups within the organisation. Limits of authority will vary depending on the responsibility and experience of the employee and the duties expected of them by the company.

Ultimately the employer is responsible for the acts or omissions of their employees, including any financial or legal repercussions. On a project, the employer has a duty to ensure that all personnel are competent to complete the tasks they are required to carry out. So it is in the employer's interests to set limits of authority appropriately.

Apprentices who are trainees will not be permitted to make decisions that are beyond their capabilities. Their authority is limited to common sense, basic health and safely matters, and those areas where they have been adequately trained.

A supervisor, with Level 3 qualifications and more training and experience than trainee staff, would have considerably more authority. Supervisors are responsible for ensuring that projects run smoothly, for developing risk assessments and method statements, and for ensuring that agreed assessments and statements are in place and adhered to. Supervisors also have financial responsibilities in ordering materials and requesting labour to respond to work programmes.

Assessment criteria

2.2 Specify organisational policies/ procedures for the handover and demonstration of electrotechnical systems, products and equipment, including requirements for confirming and recording handover

Practical completion

The point at which a construction project is virtually finished, when the last percentage of monies are paid by the client. The responsibility for insuring the construction transfers to the client and the architect or contract administrator signs the certificate of practical completion. At this point there will still be a few minor snags to sort out (for example, scratches in paintwork) or insignificant items to be completed.

ASSESSMENT GUIDANCE

On all jobs, it is good practice to walk round the site with the client, explaining the operation of the electrical equipment. When carrying out any practical work under the supervision of your assessor, tell them exactly what you are doing, in the same way.

KEY POINT

The CIBSE is the Chartered Institute of Building Services Engineers.

KEY POINT

A smaller amount of specific information is usually better received than volumes of standard information.

HANDOVER INFORMATION

Handover information needs to be delivered in order to achieve **practical completion**. As buildings have become more sophisticated, operation and maintenance documentation, for example, has become even more important. Handover information should include:

- a design philosophy and statement to say how the building has been designed and how it is to be operated (often now part of the building logbook)
- Building Regulations tests and approvals, including the sign-off documents from the Building Control officer or approved inspector
- building leakage test certificates from construction
- the energy performance certification and registration information
- electrical test information
- lighting compliance certification (eg compliance with CIBSE LG7)
- emergency lighting test certification
- fire alarm certification
- emergency voice communication (EVC) or disabled refuge system certificate
- construction plans at a scale of 1:50 or 1:100 as appropriate
- building envelope (fabric of the building) details
- structural plans with floor loadings, any specialist support information for plant, etc
- mechanical and electrical equipment layouts at a scale of 1:50 or 1:100 as appropriate
- plant and specialist area layouts at a scale of 1:50, or even 1:20, depending on complexity of the systems
- additional information relating to the cleaning, maintenance and deconstruction of any of the systems or components and any disposal requirements at the time of writing the handover information.

It is most important that the documents are up to date and specific and relevant to the building or project being handed over.

Large amounts of information need to be absorbed and understood by everyone in or allied to the construction industry. Most of this information is guidance or commentary on regulations. However, there are also laws covering electrical work and the way it is carried out, that need to be obeyed.

The law will be satisfied if the installer complies with various documents, including BS 7671 Requirements for Electrical Installations (the IET Wiring Regulations), the Institute of Engineering and Technology (IET) Guidance Notes and other related guidance. This section will help you to identify the relevant duties set out in **statute** law and commercial contract law (**civil law**) and to know how to carry out those duties.

SmartScreen Unit 303
Handout 16

Statute

A law made by Parliament as an Act of Parliament.

Civil law

Law that deals with disputes between individuals and/or organisations, in which **liability** is decided and compensation is awarded to the victim.

The law affects our everyday working lives

Assessment criteria

3.1 State the applicable health and safety requirements with regard to overseeing the work of others

3.2 State the procedures for:

■ interpreting risk assessments
■ applying method statements
■ monitoring changing conditions in the workplace
■ complying with site organisational procedures
■ managing health and safety on site
■ organising the safe and secure storage of tools and materials

5.2 Specify procedures for carrying out work activities that will:

■ maintain the safety of the work environment
■ maintain cost effectiveness
■ ensure compliance with the programmes of work

5.3 Identify the industry standards that are relevant to activities carried out during the installation of electrotechnical systems and equipment, including the current editions

ASSESSMENT GUIDANCE

You must be able to distinguish between statutory and non-statutory regulations.

STATUTORY REGULATIONS

A statutory regulation is easily distinguishable from other regulations because its full title usually contains the word 'Regulations' and always has an SI (Statutory Instrument) number.

Statutes (Acts of Parliament) are known as primary legislation. They lay down the general framework of the law, which is added to over the years through numerous regulations. Statutory regulations are known as secondary legislation.

Health and Safety at Work etc Act 1974

The Health and Safety at Work etc Act 1974 (HSW Act) is the primary legislation covering occupational health and safety in Great Britain. This Act has been amended over the years by various pieces of legislation to enforce health and safety law in Great Britain. (The Health and Safety at Work (Northern Ireland) Order 1978 applies in Northern Ireland.)

Principles of the HSW Act

All workers have a right to work in places where risks to their health and safety are properly controlled. Health and safety is about stopping anyone getting hurt at work or becoming ill through work. It sets down employers' responsibilities for health and safety, but also requires employees to help keep themselves and others safe.

The HSW Act requires employers to do the following:

■ Assess risks in the workplace and to put in place measures to prevent harm and to inform all concerned about who is responsible for the removal or reduction of the risks.
■ Consult with workers and health and safety representatives to protect employees from harm in the workplace.
■ Provide free health and safety training to enable employees to do their jobs safely.
■ Provide, free of charge, any equipment or personal protective equipment required for employees to do their jobs, and ensure it is properly maintained and remains functional.
■ Provide adequate welfare facilities such as toilets, washing facilities and drinking water.
■ Provide adequate first-aid facilities, as appropriate for the type of work being carried out.
■ Report deaths and major injuries at work to the Health and Safety Executive (HSE) Incident Contact Centre.
■ Report other injuries, diseases and dangerous incidents under the Reporting of Injuries, Diseases and Dangerous Occurrences Regulations 1995 (RIDDOR) to www.hse.gov.uk.

- Have insurance that covers employees in case they get hurt at work or become ill through work. Display a printed or electronic copy of the current insurance certificate where all employees can read it easily.

- Work with any other employers or contractors sharing the workplace or providing employees (such as agency workers), so that everyone's health and safety is protected.

The HSW Act requires employees to:

- follow the training received when using any equipment supplied by the employer

- take reasonable care of their own and other people's health and safety

- cooperate with the employer on health and safety matters

- tell someone (the employer, supervisor or health and safety representative) if they think the work or inadequate precautions are putting anyone's health and safety at serious risk.

If an employee has concerns about health and safety in the workplace:

- They should talk to their employer, supervisor or health and safety representative.

- They should seek advice from the general information about health and safety at work available on the HSE website at www.hse.gov.uk.

- If, after talking with their employer, they are still worried, they can contact their local enforcing authority for health and safety and the Employment Medical Advisory Service via the HSE website.

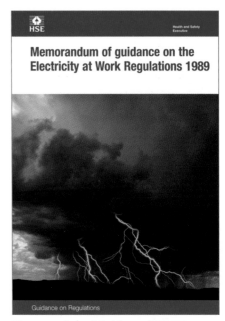

Memorandum of Guidance (HSR25)

Electricity at Work Regulations 1989

The Electricity at Work Regulations 1989 (EAWR) are made under the HSW Act. This means that breaches in the EAWR will result in action being taken under the HSW Act.

In addition to the duties imposed by the HSW Act, the EAWR impose duties on **duty holders** in respect of systems, electrical equipment and conductors, and work activities on or near electrical equipment under their control. Managers of mines and quarries are also included, despite mines and quarries having other special regulations.

The EAWR cover the principles of electrical safety that apply to work activities and systems, including anything that influences them, equipment, isolation and safety systems, and the competency of those working with electricity.

The EAWR apply beyond those situations that we traditionally associate with the dangers of electricity, including voltages outside the scope of BS 7671. The EAWR cover battery operated systems to extra high voltage transmission supplies.

KEY POINT

The EAWR apply to workplaces and not private homes, except where businesses are run from home. The HSE publication HSR25 interprets the regulations and gives guidance on how to meet the requirements.

Duty holder

Any person or organisation holding a legal duty under the Health and Safety at Work etc Act 1974.

KEY POINT

The importance of competency cannot be understated. It is possible for an individual to be responsible for causing danger to arise outside the installation he/she has designed, installed or is responsible for due to the impact of that system on other systems.

The HSE produces a memorandum of guidance (HSR25), which is free and provides guidance on how each regulation should be interpreted. Each regulation covers a specific topic, as follows:

- **Regulation 4** – systems, work activities and protective equipment
- **Regulation 5** – strength and capability of electrical equipment
- **Regulation 6** – adverse or hazardous environments
- **Regulation 7** – insulation, protection and placing of conductors
- **Regulation 8** – earthing or other suitable precautions
- **Regulation 9** – integrity of referenced conductors
- **Regulation 10** – connections
- **Regulation 11** – means for protecting from overcurrent
- **Regulation 12** – means for cutting off the supply and for isolation
- **Regulation 13** – precautions for work on equipment made dead
- **Regulation 14** – work on or near live conductors
- **Regulation 15** – working space, access and lighting
- **Regulation 16** – persons to be competent to prevent danger and injury.

Electricity Safety, Quality and Continuity Regulations 2002

The Electricity Safety, Quality and Continuity Regulations 2002 (ESQCR) impose requirements regarding the installation and use of electric lines and the apparatus of electricity suppliers, including provisions for connection with earth. The safety aspects of these regulations are administered by the HSE. All other aspects are administered by government and the industry.

The ESQCR may impose requirements, usually on supply companies or those associated with the supply of electricity, in addition to those of the EAWR. Designers of installations have a responsibility to ensure they meet the ESQCR.

Provision and Use of Work Equipment Regulations 1998

The Provision and Use of Work Equipment Regulations 1998 (PUWER) ensure that work equipment, as constructed or adapted, is suitable for the task and safe to use, taking into account all the risks of using it in a specific work environment. This applies to any equipment, machinery, appliance, apparatus, tool or installation for use at work.

For example, in order to be fit for purpose, construction site equipment needs to cope with a wide range of weather conditions including wet weather. Equipment also needs to be robust enough and sufficiently protected to withstand mechanical impact or abrasion.

It should help to reduce to as low a level as possible, the risk of electric shock through external influences, including from reduced low voltage systems such as the 110 V system used on UK construction sites. This may mean that additional protection must be provided.

Control of Substances Hazardous to Health (COSHH) Regulations 2002

The law requires employers to control substances that are hazardous to health and to protect employees and other persons from the hazards of substances used at work, by risk assessment, control of exposure, health surveillance and incident planning. The COSHH regulations contain very detailed information relating to each type of harmful substance, depending on its physical state and how it will harm individuals or groups of people.

The COSHH regulations cover:

- chemicals and products containing chemicals
- dusts, fumes, vapours and mists
- nanotechnology
- gases and asphyxiating gases
- biological agents (germs) and germs that cause diseases.

The COSHH regulations do not cover lead, asbestos or radioactive substances, which are dealt with under separate regulations.

Duties of employers

In order to reduce the exposure of workers to hazardous substances, employers are required to do the following:

- **Identify the health hazards** – using information about any chemicals or substances used in the workplace, that might be a hazard.

- **Decide how to prevent harm to health** – using risk assessment to determine who may be harmed by a chemical or substance and how someone might be harmed.

- **Provide control measures to reduce harm to health** – looking at the potential for harm and introducing protection measures, such as local exhaust ventilation (LEV), respiratory protective equipment (RPE) or personal protective equipment (PPE).

- **Make sure protective measures are used** – including ensuring that workers are informed of the risks associated with a hazard, what the protective measures are and how to comply with them, and the health and safety, and disciplinary, consequences of failing to use the measures.

- **Keep all control measures in good working order** – all measures must function correctly and any failing or potential failing must be reported and rectified. Equipment that is not in good working order must be taken out of service until a repair has taken place.

- **Provide information, instruction and training for employees and others** – this is vitally important to ensure that those who work with or are potentially exposed to harmful substances, understand what they are being exposed to, know what to do to prevent a situation getting out of control and becoming an emergency, and also know how to deal with particular situations or emergencies.

ASSESSMENT GUIDANCE

You should know what protective measures are available for common materials encountered on site.

ACTIVITY

List the items around your home that might need to be on the COSHH register if it were in a workplace. Use the internet to help you identify COSHH items.

- **Provide health monitoring in appropriate cases** – this applies to people who regularly work with or who may have been accidentally exposed to a harmful substance. If adverse effects on health become apparent, the relevant healthcare can then be undertaken.

- **Plan for emergencies** – providing an emergency contingency plan so that workers and managers know exactly what to do should the need arise.

Company name		Department/site		Assessment discussed with employees (include date)		
Potential hazard	**Potential harm and those affected**	**Existing measures**	**Additional measures required**	**Who**	**When**	**Check**
Oil-based fluid and fine particle in lathe sumps	Dermatitis Metal workers and turners	Fluid stops when machine stops 0.4 mm nitrile gloves	Improve washing facilities Apply barrier cream before and after work			
		Check oil temperature and particle concentration	Check appearance daily Fit sump thermometer			
		Change and launder overalls once a week	Keep oily rags out of pockets			

Page 1 of typical COSHH risk assessment form

Also	Action taken	Action needed		
Thorough examination and test for COSHH compliance				
Supervision	Yes			
Instruction and training	Yes			
Emergency plans	Spillage clearance	Practice		
Health surveillance		Improve record-keeping		
Monitoring				
Review date:	Review observations/actions Any significant changes:			

Page 2 of typical COSHH risk assessment form

Environment Act 1995

The Environment Agency was formed, under the Environment Act 1995, by the merger of the National Rivers Authority, Her Majesty's Inspectorate of Pollution and the local waste authorities.

The agency is responsible for waste legislation. Business waste is waste from any business or commerical activity in construction, demolition, agriculture or industry, including home businesses. Businesses have a duty of care for waste from its production through to its appropriate disposal, even when on a licensed disposal site.

The Environment Agency is also responsible for compliance with the Waste Electrical and Electronic Equipment (WEEE) Regulations 2006, under which manufacturers have to supply information relating to the decommissioning and disposal of their products.

Disability Discrimination Act 1995

The Disability Discrimination Act 1995 has now been replaced by the Equality Act 2010, but still applies in Northern Ireland. It makes it unlawful to discriminate against people because of their disabilities and requires employers to provide suitable equipment and systems for people with disabilities to do their work. However, it is still permissible for employers to have reasonable medical criteria for employing people and to expect adequate performance from those people.

Equality Act 2010

The Equality Act 2010 came into force in October 2010 and replaces all previous equality legislation in England, Scotland and Wales. It ensures that everyone has equal opportunities, regardless of age, gender, disability, race, religion/belief, pregnancy/maternity, gender/sexual orientation, marital or civil partnership status.

Manual Handling Operations Regulations 1992

The Manual Handling Operations Regulations 1992 (as amended) cover the moving of objects, but do not set specific thresholds such as weight limits for manual handling. The following points must therefore be considered:

- task
- load
- working environment
- individual capability
- other factors such as protective clothing.

The 1992 Regulations aim to protect workers from musculoskeletal disorders, which are the most common type of occupational ill health in the UK. They require control measures to be in place in the following order of importance:

1 **Avoidance** – Where possible, tasks should be designed to avoid hazardous manual handling operations so far as is reasonably practicable. This may include redesigning the task to avoid moving the load or by automating or mechanising the process to avoid the potential hazard.

2 **Assessment** – Where avoidance is not appropriate or possible, those responsible should make a suitable and sufficient assessment of any hazardous manual handling operations, to minimise risk.

3 **Reduction** – In order to reduce the risk of injury from manual handling operations, so far as is reasonably practicable, mechanical assistance should be provided, such as the use of a sack trolley or hoist. Where an aid is not reasonably practicable, there must be a practical assessment of ways to change the task, load and working environment.

Team handling/lifting may be appropriate as a group can handle or lift more than an individual. To be effective, there needs to be sufficient space for handlers to work as a group, with the appropriate equipment (for example, slings) in place and with good communication.

ACTIVITY

When using a sling to transport an oddly shaped load, why is it important to suspend the load from a point as close to the centre of gravity as possible?

The 1992 Regulations require employers to ensure that employees understand and are adequately trained in manual handling safety. Employers also have to keep records of the training carried out.

Control of Noise at Work Regulations 2005

These regulations (the Noise Regulations) require employers to eliminate or reduce risks to health and safety from noise at work. Other legislation requires users of particular types of equipment to protect workers from harm, including noise. However, the Noise Regulations look beyond equipment noise to any noise emissions that risk harm to health and safety.

Depending on the level of risk, employers should:

- take action to reduce the noise exposure
- provide employees with personal hearing protection.

Other duties under the regulations include the need to:

- make sure the legal limits on noise exposure are not exceeded
- maintain equipment provided to control noise risks, and ensure its use
- provide employees with information, instruction and training
- carry out health surveillance (monitor workers' hearing ability).

The Noise Regulations do not apply where people who are not at work are exposed to risks to their health and safety from noise related to work activities. However, the general duties in section 3 of the Health and Safety at Work etc Act 1974 may apply in such cases.

Construction (Design and Management) Regulations 2007

The Construction (Design and Management) Regulations 2007 (CDM 2007) are a key piece of construction safety legislation, which covers safe systems of work on construction sites. It makes all stakeholders accountable, including clients and designers. Its requirements are usually to be encountered in the pre-construction health and safety plan of a project.

Under the CDM 2007, some projects must be notified to the HSE in writing. This applies if:

- the project is expected to last longer than 30 days
- there will be five or more workers on site
- the project will involve more than 500 person days of construction work.

The notification must be issued on HSE form F10, which has to be displayed on the construction site.

Information needed to ensure health and safety on site

The principal contractor has responsibilities to secure health and safety on site. This involves gathering information from the contract documents, including the design specifications, and all stakeholders, including:

- the client
- the design team
- other contractors on site
- specialist contractors and consultants
- trade and contractor organisations
- equipment and material suppliers.

Additional information will also be required on special features such as:

- asbestos or other contaminants
- overhead power lines and underground services
- unusual ground conditions
- public rights of way across the site
- nearby schools, footpaths, roads or railways
- other activities going on at the site.

Managing safety on site before work starts

The principal contractor must assess access to the work site, arrange a programme of work and ensure that access arrangements are suitable for sub-contractors to use. Any specialist contractors who need to work in isolation for safety reasons (eg with chemicals or radio isotopes) should be consulted and appropriate time allowed in the programme of work for them to carry out their duties safely.

In conjunction with the sub-contractors, the principal contractor must ensure that:

- all relevant working methods have been identified and time allowed for them to be undertaken
- all equipment used on site is compatible with other operations on site at the time allocated on the programme
- all potential hazards have been considered and as many as possible eliminated (for example, painting does not need to be done at height if materials are brought to site already finished)
- all remaining risks are assessed, decisions are taken on how to control them and procedures are put in place
- everyone working on site is competent and has the right equipment to do their work

- appropriate work methods are agreed
- site deliveries are arranged at the best times to minimise safety risks to the public, taking into account neighbourhood circumstances, such as school drop-off and pick-up times
- emergency and rescue procedures for the site are in place, including any additional training required (for example, harness rescue from height).

Ensuring workers' competence and awareness of risks

Those who supervise the work must be adequately trained, have suitable experience and have access to additional guidance to ensure they are competent. They must ensure that agreed work methods are put into practice.

Workers must be competent to carry out their work, so their trade certificates and Construction Skills Certification Scheme (CSCS) card or trade skill card should be checked. They must have suitable equipment to carry out their work.

Cutting steelwork, wearing protective clothing, mask and gauntlets

It is also important to ensure that sub-contractors on site have adequate supervision and that their operatives are trained and competent in the tasks they are employed to carry out.

Before any worker starts work on site, it is important to make sure they understand the general risks on the site, are familiar with the tasks they are being employed to carry out and are aware of any additional hazards or increased risk due to local conditions. This information will usually be set out in the workers' method statement, worked through at the site induction and checked on a regular basis, with formal follow-up in toolbox talks.

ACTIVITY

Why is it important not to undertake any task unless you are totally happy with what you are doing?

Health and safety during the construction phase

During the construction phase, there must be adequate management and supervision on site to ensure all operations are carried out as safely as possible. During construction, this includes a wide range of risks. These include risks from moving goods, working at height or in confined spaces or with electricity or dangerous substances, or even risks to the public. It is likely that the greater the risk, the greater the degree of control and supervision required.

ASSESSMENT GUIDANCE

You may be asked to identify or recall materials, used by the electrician, which fall under the COSHH regulations.

It is necessary to ensure that all measures put in place are relevant and effective. It is therefore important to review procedures on a regular basis to make sure they continue to be relevant and effective as time goes by and the nature of work on site changes.

Monitoring of health and safety on site must be carried out by a trained and competent person, who is able to review the safety procedures in the context of all the changes in work and circumstances on site. The review process must also check that contractors and sub-contractors are working safely and in line with their method statements. Action must be taken to deal with those who fail to work safely. The review should also consider whether similar problems reoccur, find out why and prompt appropriate action before there is an accident. This process should include monitoring minor accidents and near misses in the accident book, as this could give early warning of the potential for a more serious incident.

Published guidance and codes of practice

Guidance on the legal requirements of certain statutory regulations and how to interpret them, and approved codes of practice (ACoPs), are published or approved by the HSE and the relevant secretary of state. These documents often contain a copy of the legislation they are giving guidance on. They have a special legal status in law. Where a duty holder does not follow the guidance or code of practice, they must show that their course of action or omission is no less safe.

L23 Manual Handling: Manual Handling Operations Regulations 1992 (as amended)

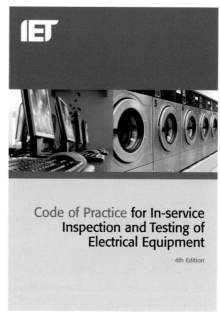

IET Code of Practice for In-service Inspection and Testing of Electrical Equipment

NON-STATUTORY REGULATIONS

Not all documents using the term 'regulations' are statutory instruments. For example, BS 7671 Requirements for Electrical Installations, commonly called the IET Wiring Regulations, are non-statutory. They form the national standard in the UK for low voltage electrical installations. They deal with the design, selection, erection, inspection and testing of electrical installations operating at a voltage up to 1000 V a.c. Work undertaken in accordance with BS 7671 is almost certain to meet the requirements of the Electricity at Work Regulations 1989. BS 7671 is supported by a number of guidance notes published by the IET.

Other **CENELEC** countries use the term 'rules' rather than 'regulations'. For example, the national wiring standard in Southern Ireland is the National Rules for Electrical Installations.

Other non-statutory rules include the rules and requirements of regulating bodies such as the NICEIC, the National Association of Professional Inspectors and Testers (NAPIT), the Electrical Contractors' Association (ECA) and trade unions such as Unite. Individuals and enterprises that want to belong to these organisations must comply with the relevant organisation's rules. In certain circumstances, compliance with such rules, such as those for competent person schemes (CPSs), assist in compliance with specific statutory regulations or requirements.

BS 7671, the IET Wiring Regulations

CENELEC

CENELEC (Comité Européen de Normalisation Électrotechnique) is the European Committee for Electrotechnical Standardization.

APPLYING STATUTORY REGULATIONS

Statutory regulations cover the whole range of areas dealt with by the law. Each set of regulations lays down the detailed requirements that have to be met to comply with a specific Act of Parliament. In construction-related industries, most legislation focuses on health and safety.

Statutory regulations place specific requirements on duty holders to do something, provide something or declare something. In the case of health and safety law, non-compliance can impact on an organisation's reputation; for example, some clients monitor HSE incidents before appointing contractors and sub-contractors. More significant failures can result in fines and for the most serious cases, prison sentences.

KEY POINT

NICEIC is the brand name of the organisation formerly known as the National Inspection Council for Electrical Installation Contracting. NICEIC is now part of the Acertiva Group Ltd.

HSE inspections

HSE inspectors routinely inspect places of work. Although it is unusual to call unannounced, they have the right to enter and inspect any workplace without giving notice.

ASSESSMENT GUIDANCE

You should be able to identify statutory and non-statutory documents.

ACTIVITY

As an apprentice, what action should you take if approached by an HSE inspector on site?

On a routine visit, rather than responding to a complaint or particular incident, inspectors inspect the workplace, work activities, and methods and procedures used for the management of health and safety, and check compliance with health and safety law relevant to the specific workplace. The inspector is empowered to talk to employees and their representatives, and take photographs and samples.

Where breaches in requirements are observed, the inspector can serve the following, depending on the severity of the breach:

- **Informal warning** – Where the breach is minor, the inspector can explain on site what is required to the duty holder (usually the employer), following up with written advice explaining best practice and the legal requirements.

- **Improvement notice** – Where the breach is more serious, the notice will state what has to be done, why and when the remedial action has to be completed by. The inspector can take further legal action if the notice is not complied with within the time period specified. The duty holder has the right to appeal to an industrial tribunal.

- **Prohibition notice** – Where there is a risk of serious personal injury, the inspector can stop the activity immediately or after a specific period. The activity cannot be resumed until remedial action has taken place. The duty holder has the right of appeal to an industrial tribunal.

- **Prosecution** – In certain circumstances, the inspector may consider prosecution necessary to punish offenders and deter other potential offenders.

In addition, particular pieces of legislation may carry specific penalties, such as unlimited fines and custodial sentences. These measures are generally taken to court by an enforcing authority. In cases of health and safety legislation violation, this is done by the HSE.

When cases go to court, specialist lawyers are called on to work through the complexities of the relevant area of law. Often the evidence of expert witnesses is sought to obtain professional opinions on the subject. The outcome will be an acquittal or a sentence with remedial/corrective action.

In general, employers are responsible for the actions of their employees and, where these result in breaches in legislation or accidents, it is the employer that is punished. However, where a breach is entirely due to an employee, the employer has the right to impose its own disciplinary processes, which, at worst, may result in a finding of gross misconduct and dismissal.

BREACHES OF NON-STATUTORY REGULATIONS

Breaches of non-statutory regulations such as British Standards (BSs) and supporting guidance, usually amount to a breach of contract and often revolve around quality. Contractual requirements for the provision of services or standards usually exceed any legal requirements. When requirements have been breached, financial compensation for damages is often sought.

Best practice that includes standards of excellence not reasonably expected in a normal contract is often referred as 'Rolls Royce standard'.

Standards required by statutory and non-statutory documents regulations

In the diagram below, showing the hierarchy of work-related legislation and guidance, it might appear that British Standards and guidance documents are less important than statutory regulations. However, the best practices contained in non-statutory documents are used as benchmarks in legal proceedings and failure to meet them can often result in a breach of legal requirements.

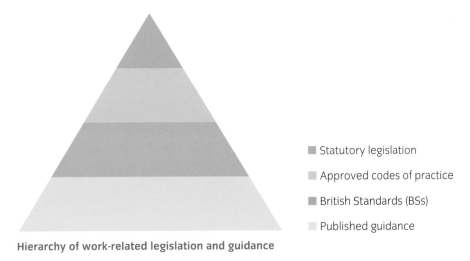

Hierarchy of work-related legislation and guidance

SmartScreen Unit 303
Worksheet 16

ACTIVITY

Compliance with BS 7671 will normally ensure meeting the requirements of which statutory regulations?

This will lead to the statutory enforcement system and criminal courts. For example, a breach in the non-statutory IET Wiring Regulations may have a significant effect on safety and ultimately result in a breach of statutory legislation.

Even if a breach in non-statutory regulations is not a legal issue, the consequences can be far-reaching. The industry may lose faith in the offending organisation's ability to perform, resulting in loss of customers and tenders. In the case of the rules of trade and similar organisations, breaches may result in expulsion from the relevant organisation.

OUTCOME 4

Understand the requirements for liaising with others when organising and overseeing work activities

You have already seen in Learning outcome 2 what methods you can use to communicate with others (pages 211–212). Now have another look at some of the methods available.

COMMUNICATING AND RESPONDING

The table below demonstrates the advantages and disadvantages of various communication methods.

Assessment criteria

4.1 Describe techniques for the communication with others for the purpose of:
- motivation
- instruction
- monitoring
- cooperation

4.4 Identify appropriate methods for communicating with and responding to others

Method	Advantage	Disadvantage
Oral in person	■ Provides an ability to judge a person's reaction and respond to it ■ Provides opportunities to confirm that instructions have been fully understood ■ Good method for motivating others ■ Good method for monitoring situations	■ Verbal agreements are not documented ■ Requires all parties to be at location
Oral by phone	■ Provides ability to judge a person's reaction (although limited) ■ Can be carried out from practically anywhere	■ No ability to judge body language ■ Conversations can be misunderstood ■ May not be in a position to answer telephone
Text	■ Good way to get small amounts of information to someone fast ■ Read receipts can be requested ■ Quick way to confirm things	■ If poorly constructed, messages can be misunderstood ■ Unable to gauge a person's reaction ■ Difficult to provide detailed information or certain graphics
Email	■ Ability to send large amounts of data and graphics ■ Read receipts can be requested ■ Quick method of communication	■ Poorly constructed emails can be misunderstood ■ Unable to gauge a person's reaction
Facsimile	■ A good way to send hand-drawn/non-electronic graphics or text	■ Sender/recipient requires fax machine or fax software
Letter	■ Good method of sending formal information which can be recorded ■ Ability to prove receipt	■ Slower method of providing information

4.2 Describe methods of determining the competence of operatives for whom you are responsible

SmartScreen Unit 303
Handout 21

KEY POINT

Competence is normally achieved through a mixture of experience and qualifications.

ASSESSMENT GUIDANCE

Do you know the roles of the ECA, NICEIC and JIB? Ask your tutor or use the internet to find out, so you can answer any questions in this area.

ASSESSMENT GUIDANCE

The achievement measurement 2 (AM2) test has a multiple-choice exam and a practical assessment. Although it has a City and Guilds number (399) it is not run by City and Guilds but by NET.

DETERMINING COMPETENCE

In order to determine whether a person is competent to carry out various tasks in the electrotechnical area, you need to understand what is required to become a qualified electrician.

To become a qualified operative involves not only passing the academic test, but being competent in the particular trade. Competence is assessed through qualifications, training and skill. The Health and Safey Executive (HSE) interpretation, given in HSR 25 Memorandum of Guidance to the Electricity at Works Regulations 1989, states:

> A person shall be regarded as competent ... where he has sufficient training and experience or knowledge and other qualities to enable him properly to assist in undertaking the measures.

In order to become a qualified electrical operative, it is necessary to have a combination of nationally recognised qualifications, with a certain amount of experience to support the qualifications. This balance will normally be assessed by potential employers. Be prepared for the fact that the majority of employers worked through full apprenticeships and may view alternatives with less enthusiasm.

There are alternative routes to qualification as a qualified electrician, which include apprenticeships for older trainees through the Sector Skills Council (SSC). The SSC is licensed by the UK government through the UK Commission for Employment and Skills (UKCES).

The principal route to becoming a qualified electrician is through the apprenticeship route. You can now take this route as an adult trainee. This involves completing a number of qualifications within the apprenticeship framework. This framework is laid down by the SSC. The apprentice needs to show that they have carried out practical training with a registered scheme and achieved 2357 (NVQ Knowledge and Practice) and, if there was no prior learning such as GCSE:

- Maths Functional Skills Level 2
- English Functional Skills Level 2
- ICT Functional Skills Level 2
- Employer's Rights and Responsibilities (ERR)
- Personal Learning and Thinking Skills (PLTS).

The alternative to the apprenticeship, which was Introduced in September 2012, is the City & Guilds 2365 Levels 2 and 3 or the older 2330 qualifications in conjunction with:

- City & Guilds 2357 Units 311–318
- 2357–399 Electrotechnical Occupational Competence (AM2).

These can be used as recognised prior learning (RPL), counting towards some of the units in the City & Guilds 2357 and hence allowing the trainee with employment that counts towards significant work experience to apply for an ECS (Electrotechnical Certification Scheme) gold card.

The ECS gold card

Areas that run competent person schemes

There are a number of areas in building services where a competent person scheme (CPS) operates, including:

- heating system installation regardless of boiler fuel
- removal or replacement of oil tanks
- installation of a new bathroom or kitchen if new plumbing is installed
- installation of additional radiators to an existing heating system
- new electrical installations in kitchens and bathrooms
- electrical work outside a house
- modification of circuits in kitchens and bathrooms (not like-for-like maintenance replacement of equipment)
- replacement of windows and door units
- replacement of flat/pitched roof coverings
- installation of fixed air-conditioning systems.

A contractor who is on a competent person scheme accreditation listing is allowed to self-certify that their work complies with Building Regulations in place of the normal Building Control notification required. Some of the areas listed, such as installation of additional radiators to an existing heating system, do not always require Building Control notification. A group of five CPS operators has formed the Electrical Safety Register. This register allows potential customers to locate registered contractors within a pre-determined radius of their own postcode. At the time of writing, the register only suggests contractors registered with two of the five operators.

Competent person schemes

A competent person scheme (CPS) is a scheme that gives its members accreditation as competent persons. Under the Electrotechnical Assessment Specification (EAS) Scheme, a competent person is defined as:

> a person, considered by the Enterprise to possess the necessary technical knowledge, skill and experience to undertake assigned electrical installation work, and to prevent danger and where appropriate injury.

ACTIVITY

The Achievement Measurement 2 test is managed by NET. Is it necessary for all NVQ candidates to take it?

ACTIVITY

What is the advantage of belonging to a competent person scheme?

All CPSs are approved and administered by the Department for Communities and Local Government (DCLG). The UK Accreditation Service (UKAS) has been appointed as the accrediting body for each of the relevant CPS operators.

A CPS operator is a body offering a self-certification scheme covering the specific area of work and meeting the requirements of UKAS accreditation requirements and Schedule 3 of the Building Regulations 2010, as amended.

In order for a CPS to become accredited, the operator must meet and maintain the requirements of UKAS BS EN 45011: 1998 or the latest equivalent. Scheme operators must assess existing members and applicants as technically competent, against National Occupational Standards (NOS) under a Minimum Technical Competence (MTC) assessment procedure, where one is in place, for the relevant type(s) of work. They must also assess the competence of prospective and existing members of the scheme to deliver compliance with the Building Regulations. New applicants must be assessed before they can be registered with the scheme.

ASSESSMENT GUIDANCE

You will need to know who operates competent persons schemes in the electrotechnical sector in the UK.

Continuous professional development presentations

Continuous professional development (CPD) seminars at a workplace or institution are common ways of giving information and know-how to the end user.

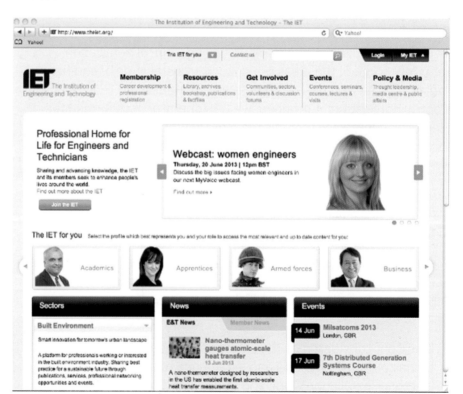

Websites such as the IET's advertise webcasts and seminars
Image courtesy of the Institution of Engineering and Technology (IET)

ACTIVITY

Visit the websites of professional institutions such as the IET, Chartered Institution of Building Services Engineers (CIBSE), Institute of Healthcare Engineering & Estate Management (IHEEM), Institution of Mechanical Engineers (IMechE) to see if there are any CPD seminars that are relevant to you. They are normally free and open to non-members.

They are often presented by an industry specialist and acknowledged expert in the field. However, seminars can also present updates to standards and upcoming changes in statutory and non-statutory regulations, all of which can have a significant effect on working methods.

Such presentations no longer need to be attended in person. They can be delivered live or recorded to an web-based audience via WebEx. After a presentation viewers can send queries to the organiser, which are then dealt with by the speaker in a short question and answer session.

WebEx seminars are often left posted on the internet so they can be referred to later. Examples are available on IET TV at www.theiet.org.

SPECIFYING ROLES

At the apprenticeship stage of your career, your role in terms of responsibility is limited. However, it must be remembered that you always have a responsibility for working safely under the Health and Safety at Work Act.

Within company policy there is likely to be a statement of the limit of authority for different individuals and groups within the organisation. Limits of authority will vary depending on the responsibility and experience of the employee and the duties expected of them by the company.

Ultimately the employer is responsible for the acts or omissions of their employees, including any financial or legal repercussions. On a project, the employer has a duty to ensure that all personnel are competent to complete the tasks they are required to carry out. So it is in the employer's interests to set limits of authority appropriately.

Apprentices who are trainees will not be permitted to make decisions that are beyond their capabilities. Their authority is limited to common sense, basic health and safely matters, and those areas where they have been adequately trained.

A supervisor, with Level 3 qualifications and more training and experience than trainee staff, would have considerably more authority. Supervisors are responsible for ensuring that projects run smoothly, for developing risk assessments and method statements, and for ensuring that agreed assessments and statements are in place and adhered to.

Supervisors also have financial responsibilities in ordering materials and requesting labour to respond to work programmes.

ASSESSMENT GUIDANCE

Remember to keep a CPD record. Many employers expect a certain number of updates each year.

Assessment criteria

4.3 Specify your role in terms of:
- responsibility for other staff
- liaison with your employer
- communication with others

Assessment criteria

4.5 Specify procedures for re-scheduling work to coordinate with changing conditions in the workplace and to coincide with other trades

PROCEDURES FOR RE-SCHEDULING WORK

You have already seen in Learning outcome 1 how work is scheduled using methods such as critical path networks (pages 190–191). Changing conditions at the workplace can lead to delays and alterations in the work programme, but good communication between the contractor and other trades can minimise these delays.

One of the most appropriate methods of overcoming changing conditions in the workplace is to hold a site meeting so the impact of the delays may be discussed by all parties concerned. If the re-scheduling of work leads to changes to the original contract, the following document would be issued.

Variation order

The purpose of a variation order is to change the conditions or clauses of an original contract due to the changing workplace or schedule. Re-scheduling work may lead to a financial burden for sub-contractors due to the need for additional material or changing labour requirements. The variation order effectively makes allowance for this change in contract.

In many cases, the changing work, materials or labour costs would be documented by the sub-contractor by issuing the contractor or client with daywork sheets.

Daywork sheets document the additional labour, materials and machinery costs involved from a variation order. Daywork sheets are generally used when the variation order costs are difficult to estimate.

Assessment criteria

4.6 Clarify organisational procedures for completing the documentation that is required during work operations

PROCEDURES FOR DOCUMENTATION

During work operations, various documents will be used within an organisation. These include:

- purchase orders
- delivery notes
- time sheets
- handover information.

Purchase orders

Purchase orders provide a method of documenting what materials or plant are purchased. By giving each order a unique number, specific materials can be accounted for to a particular project. Although many wholesalers or suppliers do not require the issue of purchase orders, it is always good practice to use them to allow the delivered items to be checked for accuracy.

Delivery notes

Delivery notes are the documents that are attached to or delivered with any delivery. It is essential that delivery notes are checked against the delivered **consignments**, wherever possible at the time of delivery. This identifies if the wrong consignment has been delivered and if items are missing. The delivery note normally requires a signature to confirm the number of items delivered.

If a problem with the delivery is discovered later, responsibility for the problem can be difficult and costly to prove. Missing components will also affect performance on site. The delay in supply of even simple products such as bolt fixings to secure the structure can affect the timing on a project badly.

Delivery notes should be checked off and processed through the supervisor. They may then be sent to the administrators for processing or direct to the contractor's quantity surveyor for costing against the project.

Consignment

A batch of goods.

ASSESSMENT GUIDANCE

You may be required to know what should be done with a delivery note and who delivery notes are handed to.

Time sheets

Time sheets are a simple but very effective tool in assigning work costs to particular jobs or budgets. They indicate where labour costs have been expended, including where additional works have been carried out and need charging to the client, which is of vital importance to profit levels.

Time sheets that break work down into specific tasks can also be used by managers and estimators to indicate areas where the estimate was totally inaccurate. This information can then be used when pricing future works.

Once submitted, a time sheet is a legal document stating the hours worked on a task or project.

KEY POINT

It is important to fill in time sheets accurately as they are a way of determining how a client should be charged for labour and what payment is owed to you.

Handover information

Handover information needs to be delivered in order to achieve **practical completion**. As buildings have become more sophisticated, operation and maintenance documentation, for example, has become even more important. Handover information should include:

- a design philosophy and statement to say how the building has been designed and how it is to be operated, often now part of the building logbook
- Building Regulations tests and approvals, including the sign-off documents from the Building Control officer or approved inspector
- building leakage test certificates from construction
- the energy performance certification and registration information

Practical completion

The point at which a construction project is virtually finished, when the last percentage of monies are paid by the client. The responsibility for insuring the construction transfers to the client and the architect or contract administrator signs the certificate of practical completion. At this point there will still be a few minor snags to sort out (eg scratches in paintwork) or insignificant items to be completed.

KEY POINT

The CIBSE is the Chartered Institute of Building Services Engineers.

 SmartScreen Unit 303
Worksheet 21

KEY POINT

A smaller amount of specific information is usually better received than volumes of standard information.

- electrical test information
- lighting compliance certification (eg compliance with CIBSE LG7)
- emergency lighting test certification
- fire alarm certification
- emergency voice communication (EVC) or disabled refuge system certificate
- construction plans at a scale of 1:50 or 1:100 as appropriate
- building envelope (fabric of the building) details
- structural plans with floor loadings, any specialist support information for plant, etc
- mechanical and electrical equipment layouts at a scale of 1:50 or 1:100 as appropriate
- plant and specialist area layouts at a scale of 1:50, or even 1:20, depending on complexity of the systems
- additional information relating to the cleaning, maintenance and deconstruction of any of the systems or components and any disposal requirements at the time of writing the handover information.

It is most important that the documents are up to date and specific and relevant to the building or project being handed over.

In order for the handover information to be complete, any instructions and manuals delivered to site along with materials should always be kept.

Understand the requirements for organising the provision and storage of resources that are required for work activities

Good storage is key to a successful contract as it can minimise loss and damage as well as reduce hazards.

Theft is commonplace on construction sites, especially at night when unattended. Providing good storage may be at a cost but it will be money well spent in the long run.

RESOURCE REQUIREMENTS AND SUITABILITY OF MATERIALS

In most cases, resources required for a particular contract will be identified during the estimating process. Work programmes will identify when, during the contract period, the resources will be required.

The essence of all estimating is to divide the installations into appropriately small units so that the cost can be estimated, with sufficient accuracy, in terms of:

- the labour (time to complete)
- the materials
- subcontracted items.

This may be achieved by dividing the estimate into smaller chunks such as control equipment at the intake position, distribution circuits and final circuits.

Assessment criteria

6.1 Interpret the installation specification and work programme to identify resource requirements for the following:
- materials
- components
- plant
- vehicles
- equipment
- labour
- tools
- measuring and test instruments

6.2 Interpret the material schedule to confirm that materials available are:
- the right type
- fit for purpose
- in the correct quantity
- suitable for work to be completed cost efficiently

SmartScreen Unit 303

Handout 28

Simple schedules

Below is an example of a simple estimating schedule.

Company name: A. N. Contracts Ltd

Client: A House builder		**Job no and description:** 1874
Return date 31 March		**Estimator:** PNB

	Item		Cat. No	quantity	Unit	Price per unit £	Discount %	Net £
1	Consumer unit 8-way c/w 2 x RCDs		12345	1	each	59.90	15	50.92
2	2-gang socket outlets		12346	20 (4 packs)	Pack of 5	18.26	30	51.13
3	2-gang surface boxes		13458	20	each	0.96	30	13.44
4	Lighting pendant sets		15677	15	each	2.98	30	31.29
5	1-gang 1-way switches		R678	15	each	2.24	30	23.52
6	1-gang 2-way switches		R679	6	each	2.24	30	9.41
7	Cable 2.5/1.5 mm^2 6242Y		C77877	160 m	100 m	62.00	40	59.52
8	Cable 1.5/1.0 mm^2 6242Y		C77978	150 m	100 m	47.00	40	42.30
9	45 A cooker switch		45689-1	1	each	6.89	30	4.83
10	Cooker switch surface box		45690-2	1	each	1.65	30	1.65
11	Cable 4/1.5 mm^2 6242Y		C8458	8	100 m	72.00	40	3.47
12	Tails 16 mm^2		66666	1 m	1 m	12.00	30	8.40
13	Recessed light fittings		6784	30	Pack of 3	27.00	30	189.00
14	Towel rail		--	1	each	24.00	30	16.80
15	Sundries							20.00

Total materials		525.68						
On cost %	30	157.71						
Total net labour	80 hours at £15.25 per hour	1220.00						
On cost %	40	488.00						
Sub total		2391.14						
VAT		478.28						
Total		2869.42						

Simple estimating schedule

Another method used to obtain a list of resources is a take-off sheet.

Area	DB reference	Drawing number	⌐○	○⌐	↓²↗	◣	⊢⊣	⊠	⋈
Ground floor	DB-2	A-2011-3-R1	12	3	18	1	6	2	3
First floor	DB-3	A-2011-4-R1	6	4	15	1	12	2	0

Sample take-off sheet

The sample take-off sheet shows the number of particular items needed for a particular area of the installation. The symbols used would be from the drawing, to minimise any confusion, and the specification would be checked for any specific manufacturer or finish required. At the bottom of the sheet, totals would indicate the quantity of each product required. The advantage of showing the quantity of items per area is that, should a contract be won, this can be used to order materials at suitable installation stages.

Other major items of specialist equipment, such as large motors and generators, will be put out for quotation to major suppliers or manufacturers.

In some situations the client may provide a Bill of Quantities. This would be a ready-made list of specific materials and would normally be made by a quantity surveyor. This would replace the take-off sheet.

It is always wise, if possible, to visit the site at this point, to see whether there are any issues that may affect the installation process. These may be issues, such as space constraints, giving delivery and storage problems (which would require a lot of labour time to be spent moving materials around from storage to site).

Equipment suitability

When the resources required are being assessed, one further consideration is the suitability of the equipment, plant or materials for the intended location. Three major factors when deciding on the suitability of the equipment are:

- environment
- user
- structure.

These three factors are covered in much greater detail in Units 304, 305 and 306 but below is a brief description for each.

Environment

The many factors to take into account when considering the environment include:

- water
- dust
- vibration
- temperature.

User

Equipment and materials must be suitable for the person(s) using them. For example, is the equipment to be used by:

- ordinary people (and if so, is the equipment suitable for use by ordinary people)
- persons with restricted movement
- children
- skilled persons only?

Structure

When considering structure, factors to be taken into account include:

- ability to support weight
- presence of combustible materials
- finish/colour
- surface- or flush-mounted structures
- fixing methods.

STORAGE

One of the most significant factors that increases the overall cost of a project, thereby reducing profit for an organisation, is poor storage.

As well as creating a health and safety issue, poor storage of equipment can also lead to items getting damaged or lost.

Type of storage

The type of storage used is very dependent on the size and layout of the project. Small works will probably not require much more storage than a van. Larger projects may require secure storage containers or specially built storage facilities within a building.

It is important for the storage facility to be secure as theft on unattended sites is commonplace.

In some situations, storage may be very limited due to the location of a building, for example in a city centre. In this situation, materials would be delivered to site on an as needed basis.

SmartScreen Unit 303
Handout 33 and Worksheet 33

ACTIVITY

What is meant by a 'just in time' delivery?

ASSESSMENT GUIDANCE

Very often, a room within a partially completed building is used by the electrical contractor for the storage of materials and tools. This has the advantage that it is close to the work area and may have security staff in constant attendance.

Storage location

The positioning of storage facilities on a construction site, for example, a storage container, will depend very much on the following:

- ease of delivery of the storage container
- ease of delivery of the equipment to be stored
- minimising distances from storage to point of use
- keeping emergency routes clear
- keeping traffic areas clear, for example lorry turning circles
- avoiding any location that is likely to be built on as part of the project or where services are to be laid
- not blocking services or systems, for example avoiding placing a container next to a wall with a flue or vent
- not blocking daylight from windows
- presence of overhead cables.

Correct location of storage facilities can lead to much more efficient working. Keeping storage clean and tidy will also lead to efficiency, as much time can be wasted looking for specific items.

WHAT YOU NOW KNOW/CAN DO

Learning outcome	Assessment criteria	Page number
1 Understand the types of technical and functional information that are available for the installation of electrotechnical systems and equipment	*The learner can:*	
	1 Specify sources of technical and functional information which apply to electrotechnical installations	178
	2 Interpret technical and functional information and data	178
	3 Identify and interpret technical and functional information relating to electrotechnical product or equipment	178
	4 Describe the work-site requirements and procedures in terms of: ■ services provision ■ ventilation provision ■ waste disposal procedures ■ equipment and material storage ■ health and safety requirements ■ access by personnel	192
	5 Identify equipment and systems that are compatible to site operations and requirements.	198

Learning outcome	Assessment criteria	Page number
2 Understand the procedures for supplying technical and functional information to relevant people	*The learner can:*	
	1 State the limits of their responsibility for supplying technical and functional information to others	200
	2 Specify organisational policies/procedures for the handover and demonstration of electrotechnical systems, products and equipment, including requirements for confirming and recording handover	216
	3 State the appropriateness of different customer relations methods and procedures	211
	4 Identify methods of providing technical and function information appropriate to the needs of others	211
	5 Explain the importance of ensuring that: ■ information provided is accurate and complete ■ information is provided clearly, courteously and professionally ■ copies of information provided are retained ■ the installation, on completion, functions in accordance with the specification, is safe and complies with industry standards.	211
	6 Describe methods for checking that relevant persons have an adequate understanding of the technical and non-technical information provided, including appropriate health and safety information.	214

Learning outcome	Assessment criteria	Page number
3 Understand the requirements for overseeing health and safety in the work environment	*The learner can:*	
	1 State the applicable health and safety requirements with regard to overseeing the work of others	218
	2 State the procedures for: ■ interpreting risk assessments ■ applying method statements ■ monitoring changing conditions in the workplace ■ complying with site organisational procedures ■ managing health and safety on site ■ organising the safe and secure storage of tools and materials.	218
4 Understand the requirements for liaising with others when organising and overseeing work activities	*The learner can:*	
	1 Describe techniques for the communication with others for the purpose of: ■ motivation ■ instruction ■ monitoring ■ cooperation	233
	2 Describe methods of determining the competence of operatives for whom they are responsible	234
	3 Specify their role in terms of: ■ responsibility for other staff ■ liaison with their employer ■ communication with others	237
	4 Identify appropriate methods for communicating with and responding to others	233
	5 Specify procedures for rescheduling work to co-ordinate with changing conditions in the workplace and to coincide with other trades	238
	6 Clarify organisational procedures for completing the documentation that is required during work operations.	238

Learning outcome	Assessment criteria	Page number
5 Understand the requirements for organising and overseeing work programmes	*The learner can:*	
	1 Describe how to plan:	178
	■ work allocations	
	■ duties of operative for whom they are responsible	
	■ coordination with other services and personnel	
	2 Specify procedures for carrying out work activities that will:	218
	■ maintain the safety of the work environment	
	■ maintain cost effectiveness	
	■ ensure compliance with the programmes of work	
	3 Identify the industry standards that are relevant to activities carried out during the installation of electrotechnical systems and equipment, including the current editions	218
	4 Identify within the scope of the work programme and operations their responsibilities	200
	5 Identify how to determine the estimated time required for the completion of the work required taking into account influential factors	178
	6 State the possible consequences of not:	178
	■ completing work within the estimated time	
	■ meeting the requirements of the programme of work	
	■ using the specified materials	
	■ installing materials and equipment as specified	
	7 Specify methods of producing and illustrating work programmes.	178

Learning outcome	Assessment criteria	Page number
6 Understand the requirements for organising the provision and storage of resources that are required for work activities	*The learner can:* **1** Interpret the installation specification and work programme to identify resource requirements for the following: ■ materials ■ components ■ plant ■ vehicles ■ equipment ■ labour ■ tools ■ measuring and test instruments	241
	2 Interpret the material schedule to confirm that materials available are: ■ the right type ■ fit for purpose ■ in the correct quantity ■ suitable for work to be completed cost efficiently	241
	3 Specify the storage and transportation requirements for all materials required in the work location	244
	4 Specify procedures to ensure the safe and effective storage of materials, tools and equipment in the work location.	244

ASSESSMENT GUIDANCE

The assessment covers the knowledge requirements of the unit and assesses all learning outcomes to verify coverage of the unit.

Part A (scenario-based assignment)

The assignment uses 23 short-answer questions and is a scenario-based, open book assignment.

You may use any resources available to you such as the internet or manufacturer information.

Part B (written examination)

The written examination is a closed book short-answer assessment and has nine questions which you need to answer in 1 hour 20 minutes.

You *must* use a pen with black or blue ink to complete your answers.

- Make sure you arrive at the designated exam room at or before the correct time.
- Write your name and candidate details at the top of your answer sheet.
- Any calculations or rough working can be done on this paper.
- You may use a scientific calculator (graphical and programmable calculators are **not** permitted).
- Mobile phones are not allowed in the examination room. They should be handed to the invigilator.
- Attempt all questions. If you find a question difficult, leave it and return to it later.

Before the assessment

- You will find some questions, starting on page 252 to test your knowledge of the learning outcomes.
- Make sure you go over these in your own time.
- Spend time on revision in the run-up to the assessment.

OUTCOME KNOWLEDGE CHECK

1   List **four** documents or sources of information, which can assist in selecting types and locations for electrical equipment and accessories to be installed.

2   Give **two** examples of how access to site by personnel and visitors may be monitored.

3   State how the following packaging may be disposed of from the site.

- Polystyrene
- Cardboard.

4   Identify the options for disposal of waste cabling and electrical accessories during refurbishment and new installation.

5   Activities such as chasing walls and cutting holes in the plasterboard ceilings of the building are to be carried out as part of the installation. Describe how suitable provision can be made for protecting the building's fixtures and fabric during these activities.

6   Acting as an apprentice on site, you are approached by the client requesting some additional work to be carried out. State:

a)   to whom the request should be passed on

b)   the document to be signed by the client before work commences

7   You are asked by your electrician to complete a day work sheet. Identify three major items which should appear on this sheet.

8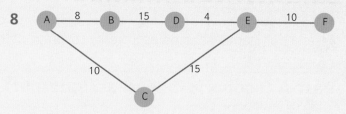

For the layout above:

a)   identify the critical path

b)   the effect of A–C increasing to 14 days

9   Name two checks to be carried out on a delivery of accessories to site.

10   Describe how you would explain the operation of an RCD and the need for quarterly testing, to a non-technical client.

11   Name three ways in which you could promote a positive 'picture' of your company on site.

12   State the main difference between a circuit diagram and a wiring diagram.

13   An electrical installation certificate is supplied for a new installation. What other forms should be supplied with the EIC?

14   An electrician wishes to list the materials he/she requires to complete an installation. They have a specification and plan of the installation. What document are they likely to use to record quantities etc?

15   State the purpose of the following:

a)   variation order

b)   site diary

c)   accident book

UNIT 304

Understanding the principles of planning and selection for the installation of electrotechnical equipment and systems in buildings, structures and the environment

In order to plan and select appropriate materials and equipment for completing electrical installations to meet a design specification, you must have a good understanding of the principles underlying the planning and design of electrotechnical equipment and systems in buildings, structures and the environment.

LEARNING OUTCOMES

There are four learning outcomes to this unit. The learner will:

1 Understand the characteristics and applications of consumer supply systems
2 Understand the principles of internal and external earthing arrangements for electrical installations for buildings, structures and the environment
3 Understand the principles for selecting cables and circuit protection devices
4 Understand the principles and procedures for selecting wiring systems, equipment and enclosures

This unit will be assessed by:

- two project-based assessments
- one written examination.

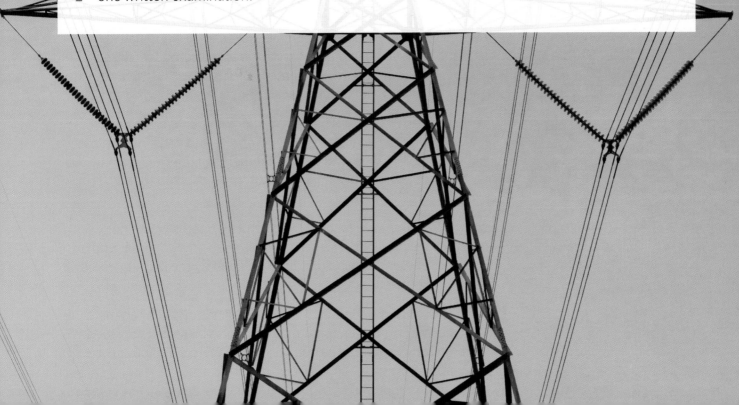

OUTCOME 1

Understand the characteristics and applications of consumer supply systems

On completion of this learning outcome, you will understand the type of electricity supply available to customers and the different arrangements for the provision of isolation and switching, overcurrent and earth fault protection.

ELECTRICITY SUPPLY ARRANGEMENTS

The majority of customers requiring an electricity supply are connected to the lower voltage networks fed from a distribution sub-station, operating at 400 V / 230 V three-phase or 230 V single-phase, and supplied via either an underground or overhead conductor.

The supply or service cable from the distribution network operator (DNO) usually terminates in a main cut-out that incorporates a protective device sized according to the load requirements. For domestic installations, this would normally be a 100 A HRC (high-rupturing capacity) fuse. For low-voltage commercial or small industrial installations, the fuse size will vary from 200 A to 400 A.

In modern domestic and small commercial installations, the electricity supply service position will be either in an external meter box that contains the fused cut-out, the electricity meter, and, more commonly now, a DNO isolator. In larger commercial or small industrial installations, the supply position will be in a dedicated switch room that contains both the DNO equipment and the main distribution board, including circuit breakers for the installation.

The DNO fused cut-out will be sealed to prevent unauthorised tampering with the fuse. When the meter tails and consumer unit are installed in accordance with the requirements of the DNO, the cut-out may be assumed to provide fault current protection up to the customer's main switch.

Note that in older installations, the electricity boards would not normally have provided a switch or isolator, so any work required on customers' meter tails would require isolation by the DNO, by removal of the fuse in the cut-out.

The meter, which may be electro-mechanical or a digital smart meter, is the property of the meter operator and will also be sealed to prevent interference by unauthorised persons. Smart meters are the latest generation of gas and electricity meters, able to provide details of how much energy is being used via a separate display located remotely from the meter and to communicate directly with the energy supplier, thus reducing the need for site visits to read the meter.

Suppliers are now beginning to provide and install a switch between the meter and the consumer unit that allows the supply to the installation to be interrupted without removal of the DNO fuse in the cut-out.

Following agreement on voltage harmonisation in 1988 by CENELEC (the European Committee for Electrotechnical Standardisation), electricity supplies within the EU are now nominally 230 V + 10% or - 6% at 50 Hz. This gives a voltage range of 216.2 V to 253.0 V for a nominal voltage of 230 V single phase and 376 V to 440 V for a nominal voltage of 400 V three phase (Appendix 2, section 14 of BS 7671).

The third stage of the harmonisation process, which will see tolerance levels moving to ± 10%, will be implemented at a later date.

Where customers use large quantities of electricity they are often supplied direct from the DNO high-voltage network, and the transformer and associated high-voltage switchgear will normally be the responsibility of the customer. The metering of such supplies will be carried out at high voltage.

Star (Y) and delta (Δ) configurations are used throughout the building services industry and, as discussed in Learning outcome 6 of Unit 309, each type of configuration has different voltage and current values. Both star and delta systems exist on three-phase systems. Star and delta are methods of connecting transformers. Most sub-station transformers used in the distribution system that delivers power to houses and commercial premises at low voltage are wound in a delta-to-star configuration. A neutral point is created in the star side of the transformer.

> ### ASSESSMENT GUIDANCE
>
> In your studies you will also come across mention of 400 V single-line supplies (although they have a two-line (phase) supply). These are not used to supply buildings but are final circuits supplying items such as welding equipment.

> ### KEY POINT
>
> Star and delta connections are not simply an option, they are used for a reason. Star transformers provide a four-wire output which provides single and three-line connections, while star mode starting for a motor reduces the starting current, before it switches into delta mode.

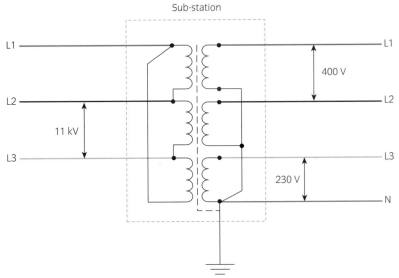

Delta/star transformer connection

Electrical systems

The Electricity at Work Regulations 1989 definition of 'system' is an electrical system in which all the electrical equipment is, or may be, electrically connected to a common source of electrical energy, and includes such source and such equipment.

An electrical system consists of:

■ the source or sources from where the electricity is generated using such prime sources as coal, gas, nuclear or renewable means of generation

■ the distribution (the means of moving the energy to where it is needed), such as cables or overhead lines

■ the installation, including the electrical wiring and switchgear in a commercial building, dwelling or shop

■ the electrical equipment, machines, IT equipment, water heaters, air conditioning etc and domestic appliances connected to the installation.

There are a number of types of electrical system that connect the distribution network to an individual installation. The type or name is determined by the earthing arrangements of the particular system and these are dealt with in Learning outcome 2.

Other considerations

Generally for new installations, a TN-C-S system is adopted.

Unless it is inappropriate for reasons of safety, the Electricity Safety, Quality and Continuity (ESQC) Regulations require the distributor to provide an earth terminal for new connections. (This is not retrospective; there are many TT installations where in the past the distributor has not provided an earth.)

Accessibility and suitability of systems

There are various considerations regarding the position of components in electrical installations.

Mains position

The position of the main incoming supply terminal (for example, fused cut out) will need to be agreed with the electrical supplier. This is necessary:

■ to provide easy access to metering equipment for meter reading

■ to facilitate installation of the supply cable and equipment.

For installations with demands exceeding 100 A, for instance, the incoming supply position will have to be determined so as to minimise voltage drop and energy loss (copper loss).

Space and access

Electrical equipment must be installed and arranged so as to provide sufficient space for:

- initial installation
- later repair and replacement
- accessibility for safe operation, inspection and testing, and maintenance.

Regulations 12 and 15 of the Electricity at Work Regulations 1989 require provision to be made for adequate space and for safe access to switches and isolators at all times. The best way to meet these requirements is at the design stage. In spite of the many pressures on space in the design of new buildings, architects have to be made aware right from the start of the space requirements for electrical equipment such as switchgear, fuse boards and busbar systems. Often the switch room or sub-station is considered as an after-thought and not allocated sufficient space.

Ideally switchgear should be housed in purpose-designed switch rooms or walk-in cupboards. Such an arrangement will:

- allocate and secure working space for the equipment
- secure the equipment against unauthorised interference or vandalism.

Switch rooms should not be used as workshops or for general storage.

Where switchgear, distribution boards, etc have to be placed in workrooms, offices or corridors, the positioning of other non-electrical equipment or office furniture must not restrict free access. Access to wall-mounted switchgear should not be obstructed, for example, by general storage. The use of appropriate floor markings (for example, yellow lines) will help emphasise the need to maintain free access.

Where access is required to the rear of a switchboard (eg to cable boxes, current transformer chambers) it must have adequate working space at both front and back. The minimum space allowance must permit unrestricted body movement. In some cases, such as when adequate headroom above switchboards is needed, for example, for access to voltage transformers (VTs), switchroom ceiling heights are important.

Electricity supply information

The Electricity Safety, Quality and Continuity (ESQC) Regulations 2002 specify safety standards aimed at protecting the public and consumers from danger. The ESQC Regulations also specify power quality and supply continuity requirements to ensure an efficient and economic electricity supply service to consumers.

ASSESSMENT GUIDANCE

In most installations, the total connected load will never be fully on. The domestic supply most people are familiar with has a 100 A supply fuse and a 40 A meter, yet if you add up all the CBs or fuse ratings, it may come to 150–200 A in total. Industrial installations may need spare capacity for future extensions.

The ESQC Regulations place requirements upon the distributor to provide suitable equipment in suitable locations; see Regulation 24 below.

Regulation 24: Equipment on a consumer's premises

1 A distributor or meter operator shall ensure that each item of his equipment which is on a consumer's premises but which is not under the control of the consumer (whether forming part of the consumer's installation or not) is:

   a) suitable for its purpose;

   b) installed and, so far as is reasonably practicable, maintained so as to prevent danger; and

   c) protected by a suitable fusible cut-out or circuit breaker which is situated as close as is reasonably practicable to the supply terminals.

2 Every circuit breaker or cut-out fuse forming part of the fusible cut-out mentioned in paragraph (1)(c) shall be enclosed in a locked or sealed container as appropriate.

3 Where they form part of his equipment which is on a consumer's premises but which is not under the control of the consumer, a distributor or meter operator (as appropriate) shall mark permanently, so as clearly to identify the polarity of each of them, the separate conductors of low-voltage electric lines which are connected to supply terminals and such markings shall be made at a point which is as close as is practicable to the supply terminals in question.

4 Unless he can reasonably conclude that it is inappropriate for reasons of safety, a distributor shall, when providing a new connection at low voltage, make available his supply neutral conductor or, if appropriate, the protective conductor of his network for connection to the protective conductor of the consumer's installation.

5 In this regulation the expression 'new connection' means the first electric line, or the replacement of an existing electric line, to one or more consumer's installations.

The designer needs to have information from the supplier in order to carry out a design. The ESQCR require the distributor to supply information on request.

Regulation 28: Information to be provided on request

A distributor shall provide, in respect of any existing or proposed consumer's installation that is connected or is to be connected to his network, to any person who can show a reasonable cause for requiring the information, a written statement of:

a) the maximum prospective short-circuit current at the supply terminals;

b) for low-voltage connections, the maximum earth loop impedance of the earth fault path outside the installation;

c) the type and rating of the distributor's protective device or devices nearest to the supply terminals;

d) the type of earthing system applicable to the connection; and

e) the information specified in regulation 27(1),

which apply, or will apply, to that installation.

In reality, the supplier will not provide the specific values for the installation in question. Suppliers are more likely to specify the maximum, depending on the system arrangement.

ARRANGEMENTS FOR ISOLATION, SWITCHING, AND OVERCURRENT AND EARTH FAULT PROTECTION

Controlling current flow

It is necessary to include devices in circuits to control current flow, that is, to switch the current on or off by making or breaking the circuit. This may be required:

- for functional purposes (to switch equipment on or off)
- for use in an emergency (switching in the event of an accident)
- so that equipment can be switched off to prevent its use and allow maintenance work to be done safely on the mechanical parts
- to isolate a circuit, installation or piece of equipment to prevent the risk of shock where exposure to electrical parts and connections is likely, for maintenance purposes.

Regulation 12 of the Electricity at Work Regulations 1989 requires that, where necessary to prevent danger, suitable means (including, where appropriate, methods of identifying circuits) must be available for:

Assessment criteria

1.2 Specify the arrangements for electrical installations and systems with regard to provision for:

- isolation and switching
- overcurrent protection
- earth fault protection

ACTIVITY

In a commercial installation, identify a device suitable for each of the following:

a) isolation of the whole installation
b) isolation of a water heater
c) isolation of a photocopier supplied through a plug and socket
d) isolation of a single lighting circuit.

- cutting off the supply of electrical energy to any electrical equipment

- the isolation of any electrical equipment.

The aim of Regulation 12 is to ensure that work can be undertaken on an electrical system without danger, in compliance with Regulation 13 (work when equipment has been made dead).

Cutting off the supply – Depending on the equipment and the circumstances, this may be no more than normal functional switching (on/off) or emergency switching by means of a stop button or a trip switch.

Isolation – This means the disconnection and separation of the electrical equipment from every source of electrical energy in such a way that this disconnection and separation is secure.

From every source of electrical energy – Many accidents occur due to a failure to isolate all sources of supply to or within equipment (eg control and auxiliary supplies, uninterruptable power supply (UPS) systems or parallel circuit arrangements giving rise to back feeds).

Secure – Security can best be achieved by locking off with a safety lock (ie a lock with a unique key). The posting of a warning notice also serves to alert others to the isolation.

Switches and isolators

Switch

A switch is a device capable of making, carrying and breaking current under normal circuit conditions. The specification of the switch may give the period of time for which it will handle overload conditions and short-circuit currents. Although a switch may be able to connect a short-circuit current, it may not be capable of breaking the circuit under such conditions.

A switch may be an isolator, though not necessarily. A switch may be used to control a piece of equipment such as a light or a motor, but it will not necessarily enable the equipment or circuit to be worked on safely.

Isolator

An isolator is a device that cuts off the installation, or part of an installation, from every source of electrical energy. This ensures that work can be carried out safely on the installation, or part of the installation, that has been isolated.

Equipment standards have particular requirements for the effectiveness and reliability of the separation, such as an air gap provided by the isolator. Isolators must also be able to be made secure in the open position so that people working on the installation are not in danger from re-closure.

ACTIVITY

Using a wholesaler's catalogue, identify switches suitable for:

a) isolation

b) mechanical maintenance

c) emergency switching

d) functional switching.

Isolators are unlikely to be used as a switch unless suitably rated. This must be checked if an isolator is to be located in an area where it may be used to disconnect loads under loaded conditions. In some situations, isolators rated at, for example, 100 A may only be capable of switching 60 A.

A common fault in industrial sites occurs when isolators are placed in locations where ordinary persons use them as a convenient switch. This causes a heat build-up on the switch contacts and eventually failure of the switch or damage to the cable where it is terminated into the isolator.

Switching for mechanical maintenance

Where plant and machinery requires work to be carried out by a non-electrically skilled person and the work does not involve any form of electrical work, a device must be provided locally to the machinery that is capable of being secured in the off position (or at least supervised) and is capable of switching full-load current.

Functional switches

Functional switches can switch on or off or control equipment; the switching may be effected by semi conductors. Examples of functional switches are light switches in a room.

Emergency switches

Emergency switching is an operation intended to remove, as quickly as possible, danger that may have occurred unexpectedly. Devices with semiconductors cannot be emergency switches. An example of an emergency switch is a stop button in a workshop.

Overcurrent protection

Each installation and every circuit within an installation must be protected against overcurrent by devices that will operate automatically to prevent injury to persons and livestock and damage to the installation, including the cables. The overcurrent devices must be of adequate **breaking capacity** and be so constructed that they will interrupt the supply without danger. Cables must be able to carry these overcurrents without damage. Overcurrents may be fault currents or overload currents.

Fault currents

Fault currents arise as a result of a fault in the cables or the equipment. There is a sudden increase in current, perhaps 10 or 20 times the cable rating, the current being limited by the **impedance** of the supply, the impedance of the cables, the impedance of the fault and the impedance of the return path. The current should be of short duration, as the overcurrent device should operate.

Breaking capacity

This is the maximum value of current that a protective device can safely interrupt.

Impedance

Impedance (symbol Z) is a measure of the opposition that a piece of electrical equipment or cable makes to the flow of electric current when a voltage is applied. Impedance is a term used for alternating current circuits. The elements of impedance are:

- resistance
- inductance
- capacitance.

Overload currents

Overload currents do not arise as a result of a fault in the cable or equipment. They arise because the current has been increased by the addition of further load. Overload protection is only required if overloading is possible. It would not be required for a circuit supplying a fixed load, although fault protection would be required.

The load on a circuit supplying, for example, a 7.2 kW shower will not increase unless the shower is replaced, when the adequacy of the circuit must be checked against the new load criteria.

A distribution circuit supplying a number of buildings could be overloaded by additional machinery being installed in one of the buildings being supplied.

Overload currents are likely to be of the order of 1.5 to 2 times the rating of the cable, whereas fault currents may be of the order of 10 to 20 times the rating.

Overloads of less than 1.2 to 1.6 times the device rating are unlikely to result in operation of the overcurrent device. Regulation 433.1 of BS 7671 requires that every circuit be designed so that small overloads of long duration are unlikely to occur.

It is usual for one device in the circuit to provide both fault protection and overload protection. A common exception is the overcurrent devices to motor circuits, where the overcurrent device at the origin of the circuit provides protection against fault currents and the motor starter will be providing protection against overload.

Protective devices

Protective devices may be one or a combination of:

- fuses
- circuit breakers (CBs)
- residual current devices (RCDs).

Fuses

Fuses have been a tried and tested method of circuit protection for many years. A fuse is a very basic protection device that melts and breaks the circuit should the current exceed the rating of the fuse. Once the fuse has 'blown' (ie the element in the fuse has melted or ruptured), it needs to be replaced.

Fuses have several ratings.

- I_n is the nominal current rating. This is the current that the fuse can carry, without disconnection, without reducing the expected life of the fuse.
- I_a is the disconnection current rating. This is the value of current that will cause the disconnection of the fuse in a given time.

- Breaking capacity (kA) rating. This is the current up to which the fuse can safely disconnect fault currents. Any fault current above this rating may cause the fuse and carrier to explode.

BS 3036 rewirable fuses

In older equipment, the fuse may be just a length of appropriate fuse wire fixed between two terminals. There are increasingly fewer of these devices around as electrical installations are rewired or updated.

One of the main problems associated with rewirable fuses is the overall lack of protection, including insufficient breaking capacity ratings. Another major problem is that the incorrect rating of wire can easily be inserted when changing the fuse, leaving the circuit underprotected.

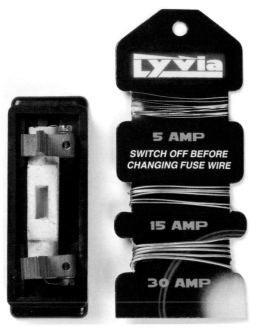

Rewirable fuse and fuse wire card, showing how wrong wire can easily be used

BS 88 fuses

These modern fuses are generally incorporated into sealed cylindrical ceramic bodies (or cartridges). If the element inside blows, the whole cartridge needs to be replaced. Although these devices have fixed time current curves, they can be configured to assist discrimination. The benefit of BS 88 and similar fuses is their simplicity and reliability, coupled with high short-circuit breaking capacity.

Within some types of BS 88 fuse, usually the bolted type, there may be more than one element. The purpose of this is to minimise the energy from a single explosion, should the fuse be subjected to high fault currents. Instead there will be several smaller explosions, allowing these devices to handle much higher fault currents of up to 80 kA.

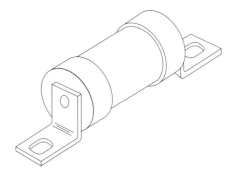

BS 88 bolted type fuse

Other BS 88 devices may be the clipped type, which do not have the two bolt tags. They are simply barrel shaped and slot into place in the carrier. They are often called cartridge fuses.

Another type of cartridge fuse is the BS 1362 plug fuse. These are fitted into 13 A plugs and are available in a range of ratings. Typical ratings are 3 A, 5 A and 13 A.

SmartScreen Unit 304
Handout 10 and Worksheet 10

Circuit breakers

Circuit breakers (CB) have several ratings.

- I_n is the nominal current rating. This is the current that the device can carry, without disconnection and without reducing the expected life of the device.

- I_a is the disconnection current. This is the value of current that will cause the disconnection of the device in a given time.

- I_{cn} is the value of fault current above which there is a danger of the device exploding or, worse, welding the contacts together.

- I_{cs} is the value of fault current that the device can handle and remain serviceable.

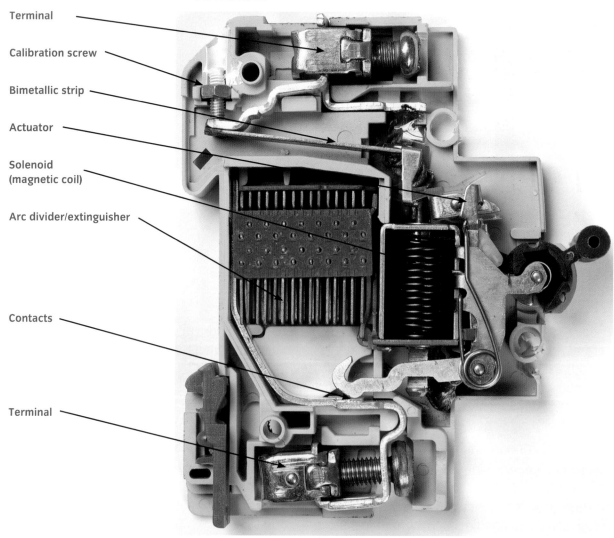

Terminal
Calibration screw
Bimetallic strip
Actuator
Solenoid (magnetic coil)
Arc divider/extinguisher
Contacts
Terminal

Section through a circuit breaker

Circuit breakers are thermomagnetic devices capable of making, carrying and interrupting currents under normal and abnormal conditions. They fall into two categories: miniature circuit breakers (MCBs), which are common in most installations for the protection of final circuits, and moulded-case circuit breakers (MCCBs), which are normally used for larger distribution circuits.

Both types work on the same principle. They have a magnetic trip and an overload trip, which is usually a bimetallic strip. If a CB is subjected to overload current, the bimetallic strip bends due to the heating effect of the overcurrent. The bent strip eventually trips the switch, although this can take considerable time, depending on the level of overload.

Miniature circuit breakers (MCBs)

These thermomagnetic devices have different characteristics, depending on their manufacture. They generally have a lower prospective short-circuit current rating than a high-rupturing capacity (HRC) fuse, ranging from approximately 6 kA to 10 kA. Specialist units are available for higher values.

The operating characteristics of MCBs can be shown in graphical form by a time–current curve.

ACTIVITY

Circuit breakers are often described as being 'trip free'. What does this mean?

KEY POINT

BS 7671 refers to both miniature circuit breakers (MCB) and moulded case circuit breakers (MCCB) as circuit breakers (CB).

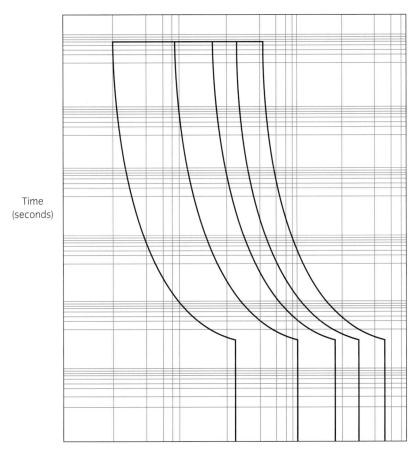

Time (seconds)

Prospective fault current
(amperes)

Sample time–current characteristic graph which is found in Appendix 3 of BS 7671

MCBs are generally faster acting than the standard curve in BS 88 fuses. A CB has a curve, then a straight line, whereas the BS 88 fuse is fully curved. This demonstrates the two tripping mechanisms in a CB. The magnetic trip is represented by the straight line on the graph, indicating that a predetermined value of fault current will disconnect the device rapidly. The curve represents the device's thermal mechanism. Like a fuse, the thermal mechanism reacts within a time specific to the overload current. The bigger the overload, the faster the reaction.

Moulded-case circuit breakers (MCCBs)

Although moulded case circuit breakers (MCCBs) work on the same principles as MCBs, the moulded case construction and physical size of MCCBs gives them much higher breaking capacity ratings than those of MCBs. Many MCCBs have adjustable current settings.

Residual current devices (RCDs) and residual current circuit breakers with overload (RCBOs)

Residual current devices (RCDs) operate by monitoring the current in both the line and neutral conductors of a circuit. If the circuit is healthy with no earth faults, the toroidal core inside the device remains balanced with no magnetic flux flow. If a residual earth fault occurs in the circuit, slightly more current flows in the line conductor compared to the neutral. If this imbalance exceeds the residual current setting of the device, the flux flowing in the core is sensed by the sensing coil, which induces a current to a solenoid, tripping the device.

Internal circuit diagram for an RCD

Residual current breakers with overload (RCBOs) combine an overcurrent protective device with a RCD in the body of the CB.

Unlike CBs, RCDs and RCBOs have a test button, which should be pressed at very regular intervals to keep the mechanical parts working effectively. If the mechanical components in a CB stick, there is not much concern as the energy needed to trip a CB is large enough to unstick any seized parts. As RCDs and RCBOs operate under earth fault conditions, with relatively small residual currents, there may not be enough energy to free any seized parts.

Application of protective devices

BS 3036 rewireable fuses

Unlike most other protective devices, the BS 3036 fuse arrangement does not have a very accurate operating time or current as it is dependent upon factors such as age, level of oxidation on the element and how it has been installed (eg whether it was badly tightened, open to air movement).

The lack of reliability of these fuses is a concern to designers and duty holders. Due to the lack of sensitivity, special factors have been applied to Appendix 4 of BS 7671 to account for these fuses. The rating factor to be applied (C_f) is 0.725.

RCBO to BS EN 61009

A range of BS 3036 rewireable fuses: 5 A (white), 15 A (blue) and 20 A (red)

BS 88 fuses

High-rupturing capacity (HRC) or high-breaking capacity (HBC) fuses are common in many industrial installations. They are also very common in switch fuses or fused switches controlling specific items of equipment. They are particularly suited to installations with a high prospective fault current (I_{pf}) as they have breaking capacities of up to 80 kA. BS 88 fuses come in two categories:

- gG for general circuit applications, where high inrush currents are not expected

- gM for motor-rated circuits or similar, where high inrush currents are expected.

ACTIVITY

What colour identification is used for a 45 A rewireable fuse?

ASSESSMENT GUIDANCE

In a fused switch the fuses are mounted on the moving contacts. In a switch fuse, the fuse and switch are in series and the fuse does not move. Remember to connect the supply to the correct terminals, otherwise the fuses will be live even when switched off.

MCBs

There are three common types of MCB: Type B, Type C and Type D. The difference between the devices is the value of current (I_a) at which the magnetic part of the device trips. The different types are selected to suit loads where particular inrush currents are expected.

Type B trips between three and five times the rated current (3 to $5 \times I_n$). These MCBs are normally used for domestic circuits and commercial applications where there is no inrush current to cause it to trip. For example, the magnetic tripping current in a 32 A Type B CB could be 160 A. So $I_a = 5 \times I_n$. These MCBs are used where maximum protection is required and therefore should be the choice for general socket-outlet applications.

Type C trips between five and ten times the rated current (5 to $10 \times I_n$). These MCBs are normally used for commercial applications where there are small to medium motors or fluorescent luminaires and where there is some inrush current that would cause the CB to trip. For example, the magnetic tripping current in a 32 A Type C CB could be 320 A. So $I_a = 10 \times I_n$.

Type D trips between ten and twenty times the rated current (10 to $20 \times I_n$). These MCBs are for specific industrial applications where there are large inrushes of current for industrial motors, x-ray units, welding equipment, etc. For example, the magnetic trip in a 32 A Type D CB could be 640 A. So $I_a = 20 \times I_n$.

MCCBs

MCCBs are available in various ranges. Lower-cost simpler versions are thermomagnetic with no adjustment. Other devices have electronic trip units and sensitivity settings or the ability to be de-rated.

Most MCCBs are used on larger circuits or distribution circuits where larger prospective short circuits are likely but the flexibility of an electronic trip is also required.

Breaking capacities

As previously mentioned in this learning outcome, one of the greatest considerations when selecting a protective device for any particular installation is the suitability of that device to disconnect a fault current safely. If a fault current exceeds the breaking capacity of a protective device, the device may:

- explode, causing a risk of fire or burns
- damage the internal components, making the device inoperable, in the case of CBs
- weld contacts together, so the device will not interrupt the fault current.

It is essential that designers of electrical installations select protective devices that have a rated short-circuit capacity or breaking capacity greater than the prospective fault current (I_{pf}) for that part of the electrical installation.

The table below shows the rated short circuit capacities for commonly used protective devices.

▼ **Table 7.2.7(i)**   Rated short-circuit capacities

Device type	Device designation	Rated short-circuit capacity (kA)	
Semi-enclosed fuse to BS 3036 with category of duty	S1A S2A S4A	1 2 4	
Cartridge fuse to BS 1361 type I type II		16.5 33.0	
General purpose fuse to BS 88-2		50 at 415 V	
BS 88-3 type I type II		16 31.5	
General purpose fuse to BS 88-6		16.5 at 240 V 80 at 415 V	
Circuit-breakers to BS 3871 (replaced by BS EN 60898)	M1 M1.5 M3 M4.5 M6 M9	1 1.5 3 4.5 6 9	
Circuit-breakers to BS EN 60898* and RCBOs to BS EN 61009		I_{cn} 1.5 3.0 6 10 15 20 25	I_{cs} (1.5) (3.0) (6.0) (7.5) (7.5) (10.0) (12.5)

* Two short-circuit capacities are defined in BS EN 60898 and BS EN 61009:

I_{cn} the rated short-circuit capacity (marked on the device).
I_{cs} the in-service short-circuit capacity.

Rated short circuit capacities of protective devices (from the On-Site Guide, IET)

Understand the principles of internal and external earthing arrangements for electrical installations for buildings, structures and the environment

Assessment criteria

2.1 Explain the key principles relating to earthing and bonding

ASSESSMENT GUIDANCE

Earthing provides a low impedance path to quickly discharge any fault current. Bonding provides an equal potential zone.

Exposed conductive parts

Conductive parts of equipment, such as metal casings, that can be touched and are not normally live but can become live when basic insulation fails.

Extraneous conductive parts

Conductive parts (eg metal, gas and water pipes) liable to introduce a potential (generally earth potential) and not forming part of the exposed metallic parts of the building's structure. These are then connected to the main earthing terminal, which in turn forms a circuit to the general mass of earth.

EARTHING AND BONDING

Regulation 8 of the Electricity at Work Regulations 1989 requires that:

'Precautions shall be taken, either by earthing or by other suitable means, to prevent danger arising when any conductor (other than a circuit conductor) which may reasonably foreseeably become charged as a result of either the use of a system, or a fault in a system, becomes so charged...'

Earthing and equipotential bonding, when used in combination and coupled with automatic disconnection of the supply, are the most common methods of providing protection to satisfy Regulation 8.

The aim is to create an earthed equipotential zone that will minimise the potential difference to which a person within the zone may be subjected in the event of an earth fault. Rapid disconnection (eg by a fuse or circuit breaker) should occur, in which case the resultant limitation of energy also forms part of the total protective measures. This is achieved by bonding together **exposed conductive parts** and **extraneous conductive** parts.

The effect of electric shock depends on time as well as the value of the current (ie the energy). It is possible for a very high voltage to be present but with a source impedance so high that the current, which would flow through a person, causes no danger (eg static electricity). It is also possible for the current to be relatively high but to persist for only a short time (eg an electric fence) without risk of injury.

Earthing arrangements of electrical systems

SmartScreen Unit 304
Handout 2

You need to be familiar with both the conductor arrangements and the earthing arrangements adopted by common supply systems.

The systems commonly adopted in the United Kingdom are TN-S, TN-C-S and TT, as shown in the diagrams below and in the following pages.

The first letter of the system type describes the earthing of the source:

- T for 'terre' (French for 'earth') when there is an earth connection of the neutral point

- I for 'isolated' when the source is not earthed.

The second group of letters describes the nature of the installation connection with earth via protective conductors:

- N-C indicates that the neutral and protective conductors are combined

- N-S indicates that the neutral and protective conductors are separated

- T is used when the installation has its own earth connection

- N-C-S indicates that the neutral and protective conductors are combined in the supplier's network and separated within the installation.

TN-S earthing systems

TN-S systems often use the lead sheath of the cable as the protective conductor. Although many TN-S arrangements still exist, the savings made by installing or maintaining one fewer conductor is the reason for electricity distribution companies converting TN-S distribution systems to TN-C-S.

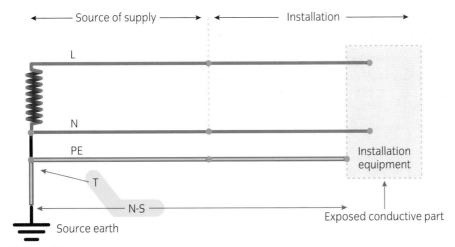

TN-S earthing system

TN-S earthing systems are often adopted for large installations supplied by their own transformer. They are also found in old (circa pre-1960) domestic installations. The diagram on the next page shows a domestic TN-S supply position arrangement.

ACTIVITY

Look up conductor arrangements and types of system earthing in Regulation 312 of BS 7671. What do the letters PME stand for?

KEY POINT

In TN-S systems, the neutral and protective conductors are separated at the system supply, for example, from the transformer, to and throughout the installation.

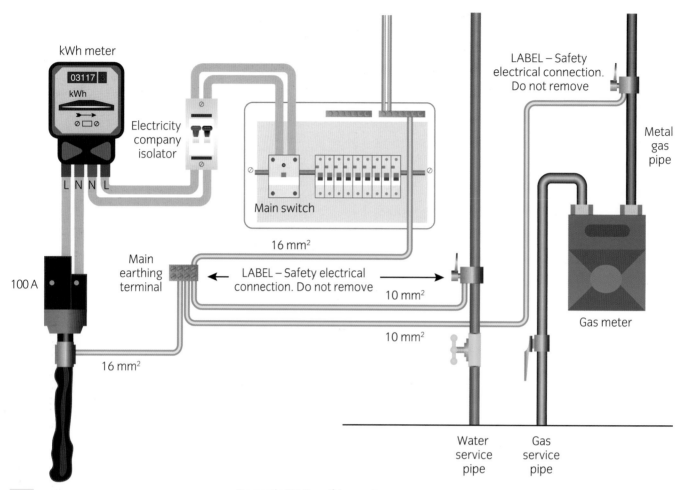

Domestic TN-S earthing system

PEN

Protective earthed neutral. The PEN conductor is both a live conductor and a combined earthing conductor.

PME

Protective multiple earthing. This term is commonly used to describe a TN-C-S system.

TN-C-S earthing systems

TN-C-S is the most common system. It is used in the great majority of installations supplied from the public supply system, particularly new installations.

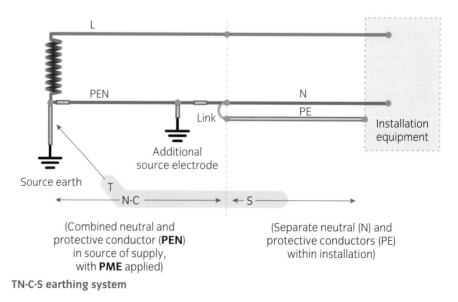

(Combined neutral and protective conductor (**PEN**) in source of supply, with **PME** applied)

(Separate neutral (N) and protective conductors (PE) within installation)

TN-C-S earthing system

Most electricity distribution systems in the UK have a combined neutral and protective conductor, called a PEN conductor. Such systems are required by the Electricity Safety, Quality and Continuity (ESQC) Regulations to be multiply earthed to provide protection to the user in the event of one or more earth connections failing. Electricity suppliers (distribution network operators or DNOs) will usually offer protective multiple earthing (PME) to someone requesting a new supply.

The Electricity Safety, Quality and Continuity (ESQC) Regulations prohibit the use of protective multiple earthing in caravans and boats. PME is also prohibited in agricultural and horticultural installations by Section 705 of BS 7671.

If the PEN conductor supplying the caravan or boat goes open circuit (breaks), all the metalwork connected to the earthing terminal may rise in potential relative to true earth. This could be a particular risk, for instance, for a person standing outside a metal caravan with bare feet in contact with wet grass.

> **KEY POINT**
>
> TN-C-S system: the neutral and protective conductors are combined in the supplier's network then separated at the supplier's fused cut-out. They remain separated throughout the installation.

> **ACTIVITY**
>
> Look at the supply position in your home. What system is it part of?

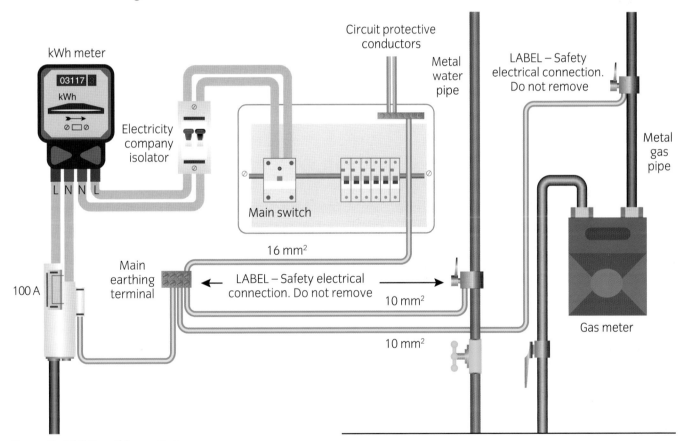

Domestic TN-C-S earthing system

Load current

The amount of current an item of equipment or a circuit draws under full load conditions.

TT earthing systems

TT systems are common in rural areas, particularly for farms and similar premises.

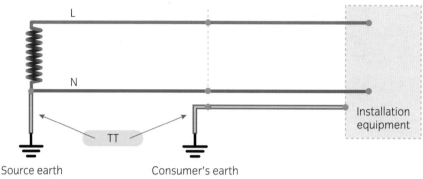

TT earthing system

Electricity suppliers will often refuse to provide protective multiple earthing (PME) to farms. This is because of the difficulty of bonding all connections with earth together in a farm and the risks of an open circuit PEN conductor. The installation must then have its own earth, not connected to the supply, and the installation becomes TT.

The connection to earth for an installation forming part of a TT system will use an earth electrode as a means of earthing. A complete list of what is accepted as an earth electrode can be found in Chapter 54 of BS 7671.

TN-C earthing systems

TN-C systems (within buildings) require an exemption from the Electricity Safety, Quality and Continuity (ESQC) Regulations and are, as a result, generally not permitted in the UK. The problem is that normal **load currents** will flow in earth conductors and all metal connected with the earth. This can cause corrosion and significant interference with information technology (IT) systems, particularly communications.

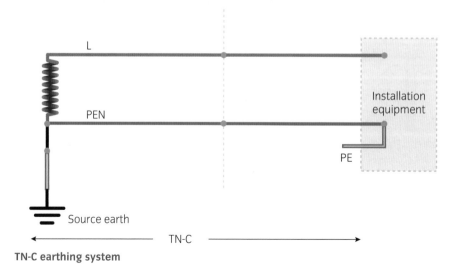

TN-C earthing system

IT earthing system

Isolated earthing (IT) is used in parts of electrical installations where interruptions to the supply can have serious consequences, for example, life-support systems in a hospital operating theatre. There is monitoring of the installation for faults without automatic disconnection. This monitoring is by the use of residual current monitors (**RCM**s) or insulation monitoring devices (**IMD**s).

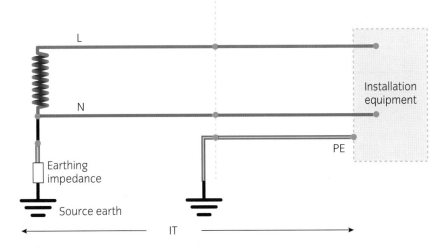

IT earthing system

RCM

Residual current monitor. This is much like an RCD but does not disconnect the circuit. Instead it gives a warning that a fault is present.

IMD

Insulation monitoring device. An end-of-line resistor together with a fire detection system is an example.

SmartScreen Unit 304
Worksheet 2

PROTECTION OF ELECTRICAL SYSTEMS

Electric shock

Electric shocks may arise either by direct contact with a live part or indirectly by contact with an exposed conductive part (eg a metal equipment case) that has become live as a result of a fault condition.

Chapter 41 of BS 7671 details the requirements for protection against electric shock and provides many ways to provide protection; these are known as 'protective measures' (see the table on page 276). Each protective measure comprises two protective provisions.

Protective provisions

BS 7671 requires two lines of defence against electric shock. These are:

- a basic protective provision (eg basic insulation of live parts)
- a fault protective provision (eg automatic disconnection of supply).

Assessment criteria

2.2 Explain the key principles relating to the protection of electrical systems

SmartScreen Unit 304
Handout 6 and Worksheet 6

KEY POINT

Protection against electric shock is one of the fundamental principles of BS 7671. Look up Chapter 13 and Regulation 131.2.

The combination of the two is a protective measure.

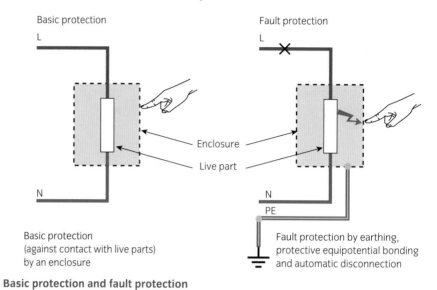

Basic protection
(against contact with live parts)
by an enclosure

Fault protection by earthing,
protective equipotential bonding
and automatic disconnection

Basic protection and fault protection

ACTIVITY

Double-insulated equipment does not have an earth connection and must not be connected to earth. Name three power tools that use double-insulated protection.

Protective measures

Protective measures	Protective provisions	
Relevant BS 7671 sections shown in brackets	**Basic protective provision**	**Fault protective provision**
Automatic disconnection of supply (411)	Insulation of live parts Barriers or enclosures	Protective earthing Automatic disconnection Protective bonding
Double insulation (412)	Basic insulation	Supplementary insulation
Reinforced insulation (412)	Reinforced insulation	Reinforced insulation
Electrical separation for one item of equipment (413)	Insulation of live parts	One item of equipment, simple separation from other circuits and earth
Extra-low voltage (SELV and PELV) (414)	Limitation of voltage, protective separation, basic insulation	
For supervised installations		
Non-conducting location (418.1)	Insulation of live parts, barriers or enclosures	No protective conductor; insulating floor and walls, spacings/obstacles between exposed conductive parts and extraneous conductive parts
Earth-free local equipotential bonding (418.2)	Insulation of live parts, barriers or enclosures	Protective bonding, notices, etc
Electrical separation with more than one item of equipment (418.3)	Insulation of live parts	Simple separation from other circuits and earth, separated protective bonding, etc

Protective measure: automatic disconnection of supply

The protective measure automatic disconnection of supply is the most commonly used protective measure in electrical installations. It requires:

1 basic protection provided by insulation of live parts or by barriers or enclosures

2 fault protection provided by:

a) earthing

b) protective equipotential bonding

c) automatic disconnection in case of a fault.

Maximum disconnection times

Chapter 41 of BS 7671 sets maximum disconnection times for earth faults that, if met, will result in the circuit meeting the fault protection requirements for automatic disconnection of supply. The table below, from BS 7671, gives disconnection times for all circuits rated up to and including 32 A.

System	$50 V < U_o \leqslant 120 V$ seconds		$120 V < U_o \leqslant 230 V$ seconds		$230 V < U_o \leqslant 400 V$ seconds		$U_o > 400 V$ seconds	
	a.c.	d.c.	a.c.	d.c.	a.c.	d.c.	a.c.	d.c.
TN	0.8	see note	0.4	5	0.2	0.4	0.1	0.1
TT	0.3	see note	0.2	0.4	0.07	0.2	0.04	0.1

Maximum disconnection times for TN and TT systems (Table 41.1 of BS 7671:2008 Requirements for Electrical Installations (the IET Wiring Regulations 17th edition) (2011))

Where, in a TT system, disconnection is achieved by an overcurrent protective device and protective equipotential bonding is connected to all the extraneous conductive parts within the installation in accordance with Regulation 411.3.1.2, the maximum disconnection times applicable to a TN system may be used.

■ U_o is the nominal a.c. rms or d.c. line voltage to Earth.

Where compliance with this Regulation is provided by an RCD, the disconnection times in accordance with Table 41.1 relate to prospective residual fault currents significantly higher than the rated residual operating current of the RCD.

Note: disconnection is not required for protection against electric shock but may be required for other reasons, such as protection against thermal effects.

BS 7671 relaxes the disconnection time to:

- 5 seconds for:
  - distribution circuits within installations forming part of a TN system
  - final circuits exceeding 32 A within installations forming part of a TN system
- 1 second for:
  - distribution circuits within installations forming part of a TT system
  - final circuits exceeding 32 A within installations forming part of a TT system.

Fault current required

Protective devices require a particular amount of fault current in order to disconnect in the required time. The higher the fault current is, the quicker the disconnection time.

If the minimum value of fault current is not reached during a fault to earth, the circuit will not disconnect effectively. For a circuit to disconnect effectively, the maximum earth fault loop impedance should not be exceeded as:

$$Z_s \leqslant \frac{U_o}{I_a}$$

where:

Z_s is the total earth fault loop impedance of the supply and installation

U_o is the nominal voltage to earth (voltage variations may need to be taken into account)

I_a is the value of current required to cause the protective device to disconnect in the specified time.

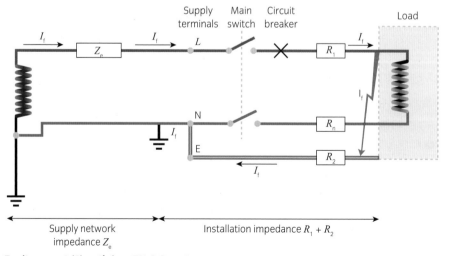

Fault current (I_f) path in a TN-C-S system

As the total earth fault loop impedance directly affects the value of earth fault current, it is the responsibility of the designer to select the parts of the earth fault path, within the installation, to ensure disconnection times are met. This includes the values of resistance $R_1 + R_2$ for any distribution circuit and to the extremities of all final circuits.

Remember:

$$I_f = \frac{U_o}{Z_e + (R_1 + R_2)}$$

where:

I_f is the fault current

U_o is the nominal a.c. rms line voltage to earth

Z_e is that part of the earth fault loop impedance which is external to the installation

R_1 is the resistance of the line conductor of the circuit at conductor operating temperature

R_2 is the resistance of the circuit protective conductor of the circuit at conductor operating temperature.

OPERATING PRINCIPLES, APPLICATIONS AND LIMITATIONS OF PROTECTIVE DEVICES

Assessment criteria

2.3 Explain the operating principles, applications and limitations of protective devices

We have already looked at how protective devices operate in Learning outcome 1. Here we look at the principles of the causes of operation.

An electrical circuit may be subjected to various abnormal conditions which, if allowed to continue unchecked, may result in cable or equipment damage. These are:

- **Overload:** Under an overload condition the circuit conductors are carrying more current that the manufacturer's design specification. If allowed to continue indefinitely, the conductors will become very hot ($P = I^2 \times R$), causing deterioration of the electrical insulation surrounding the live conductors. This may eventually lead to breakdown of the insulation and a short circuit.

- **Short circuit**: In the case of a short circuit, a breakdown of the electrical insulation has already occurred, leading to bridging or short-circuiting between conductors (line-to-line or line-to-neutral). Since the path taken by the fault current under a short-circuit condition will be of a very low resistance, the current magnitude will be very high and liable to cause severe damage. It is therefore important that short circuit faults are disconnected as quickly as possible before any cables or equipment become damaged, with the consequent risk of fire or explosion.

- **Earth fault**: In the case of an earth fault, exposed conductive parts rise to the same potential as the line conductor. This gives an immediate risk to persons or livestock in contact with exposed parts so it is important that earth faults are disconnected as quickly as possible in accordance with BS 7671.

> **To summarise**
>
> - an overload is a long-term condition
> - a short-circuit is a short-term condition

Discrimination of protective devices

There will usually be several stages of protection in large systems. It is therefore necessary to ensure that series protective devices discriminate properly in order to avoid unnecessary disconnection of parts of the system not directly associated with the fault or overload. The system designer should therefore pay particular attention to selecting the appropriate type and rating of protective devices. This requires knowledge of the protective device performance characteristics, obtainable from manufacturers' published data sheets.

Any circuit protective device must have a fault rating at least as great as the fault level at its point of installation in a system.

If there is a fault in an installation, ideally only the equipment or cable that is faulty should be disconnected so that the remainder of the installation can continue to operate normally. It may be necessary to ensure continued functioning of the unaffected parts of the installation to prevent inconvenience or to prevent danger, such as a fault in a circuit causing lighting to be lost.

Division of an installation

Section 314 of BS 7671 gives reasons why installations should be divided into circuits. These include:

- to avoid danger
- to minimise inconvenience in the event of a fault.

There are further reasons, including:

- to allow inspection and testing
- to reduce unwanted tripping of RCDs
- to reduce electromagnetic interference.

ACTIVITY

When deciding the rating of similar devices to provide discrimination, a factor of 2 is often used between the device closest to the load and the device supplying it. If the CB closest to the load is 32 A, what rating of CB should be used to supply it?

By dividing the installation into circuits, including distribution circuits, we can minimise the chance of losing more than one section of an installation.

Discrimination between devices

Effective discrimination is achieved when a designer ensures that local protective devices disconnect before others located closer to the origin of the installation do. Discrimination is required:

- under normal load conditions
- under overload conditions.

In the event of a fault, discrimination is desirable, but if no danger and no inconvenience arise it may not be necessary.

Fuse-to-fuse discrimination

Two fuses in series will discriminate between one another under overload and fault conditions if the maximum pre-arcing characteristic of the upstream device exceeds the maximum operating characteristic of the downstream device.

Discrimination will be achieved for fuses if an upstream device is more than twice the rating of any downstream device. For example, if the upstream fuse (A) has a rating of 80 A and the local downstream fuse (B) is 32 A, discrimination would be achieved as neither of the characteristic curves cross one another.

ASSESSMENT GUIDANCE

When a fuse blows or a circuit breaker trips, the action is divided into two parts. The pre-arcing energy occurs before the element melts (or the contacts open). The remainder of the energy let through occurs during the arcing period.

ASSESSMENT GUIDANCE

It is obviously possible to put all lights on one circuit in a small house, but it would be dangerous if that circuit trips out. Loads are normally split by floor or half and half between floors.

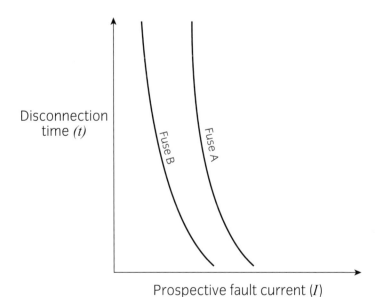

Fuse characteristics

KEY POINT

The pre-arcing time of a device is the time taken for a device to react to an overcurrent, up to the point where the device breaks and arcs.

The maximum operating characteristic is the time a device takes to complete the operation of disconnection – including pre-arcing time, arcing and disconnecting – leaving a gap big enough to prevent current flow. Manufacturers provide detailed graphs showing pre-arcing characteristics. Graphs in BS 7671 only show maximum operating characteristics.

Circuit breaker to circuit breaker

Circuit breaker characteristics are different to those of fuses. The circuit breaker has two characteristic features:

- a thermal characteristic similar to a fuse
- a magnetic characteristic when at a specified current that causes the circuit breaker to operate instantaneously (see diagram below).

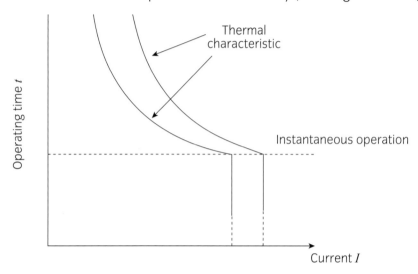

Circuit breaker characteristics

If the fault current exceeds the instantaneous operating current of the upstream device, there may not be discrimination if the two devices have the same frame type, whatever their ratings, as both devices may operate at the same time.

Manufacturers will provide information on circuit-breaker discrimination. It will normally be achieved by selecting different frame types for the upstream and downstream circuit breakers.

In some situations, a local circuit breaker may be of a different type to one installed nearer the origin. Even though the local device is only rated at 32 A (type D) and the distribution device is rated at 100 A (type B), discrimination will not be achieved.

Looking at Appendix 3 of BS 7671, we can see that a 32 A type D circuit breaker will disconnect at a current of 640 A whereas a 100 A type B will disconnect at 500 A fault current. This means that the 100 A device will disconnect before the 32 A device.

EARTH FAULT LOOP IMPEDANCE AND PROTECTIVE MULTIPLE EARTHING

Earth fault loop impedance

The earthing system adopted will determine the **earth fault loop impedance**, and this will determine the method of protection against electric shock.

Assessment criteria

2.4 Specify what is meant by the terms relating to earthing and the function of earth protection:

- earth fault loop impedance
- protective multiple earthing (PME)

Earth fault loop impedance

The impedance of the earth fault current loop starting and ending at the point of earth fault. This impedance is denoted by the symbol Z_s.

- TN-C-S systems tend to have low earth fault loop impedances external to the installation, of the order of $0.35\,\Omega$.

- TN-S systems tend to have higher earth fault loop impedances compared to TN-C-S systems. The typical maximum declared value is $0.8\,\Omega$.

- When TT systems are adopted, the resistance of the installation earth will be high (of the order of a $100\,\Omega$). This means that residual current devices (RCDs) will need to be adopted for protection against electric shock as they operate at much lower earth fault currents than standard protective devices.

The earth fault loop comprises the following, starting at the point of fault:

- the circuit protective conductor

- the consumer's earthing terminal and earthing conductor

- (for TN systems) the metallic return path

- (for TT and IT systems) the earth return path

- the path through the earthed neutral point of the transformer

- the transformer winding

- the line conductor from the transformer to the point of fault.

ACTIVITY

Draw the earth fault loop path for a fault on a TT installation.

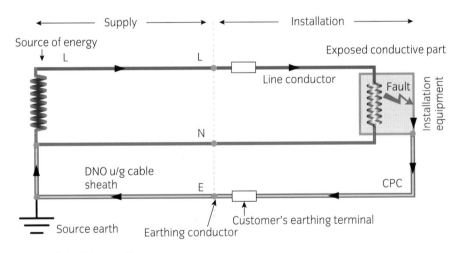

Earth fault loop

◄ Earth fault loop path

Earth fault loop

Protective multiple earthing (TN-C-S)

With this arrangement the neutral and protective functions are combined in a single conductor in part of the system. While separate N and E conductors are employed on the consumer's installation, at the main earthing terminal, a connection is made to the neutral conductor of the incoming supply cable. The neutral and protective functions are then combined as one conductor, normally the metallic cable sheath/armouring, back to the local distribution sub-station.

In addition the distribution network operator (DNO) will place earth electrodes at strategic intervals along the length of the distributor cable; this earthing arrangement is known as protective multiple earthing (PME) and is the one now offered for new installations.

PEN

Protective earthed neutral. The PEN conductor is both a live conductor and a combined earthing conductor.

PME

Protective multiple earthing. This term is commonly used to describe a TN-C-S system.

ASSESSMENT GUIDANCE

The TN-C-S system is not suitable for all installations, caravan parks and petrol stations being two of these. Where an installation is supplied by TN-C-S it must not be extended outside the building to an outbuilding, but must be converted to TT for that part of the installation.

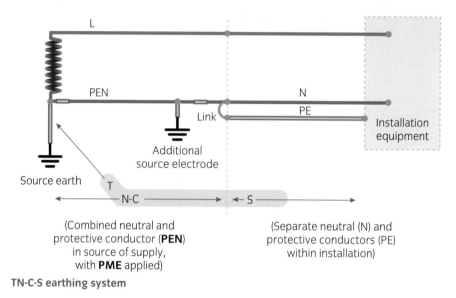

(Combined neutral and protective conductor (**PEN**) in source of supply, with **PME** applied)

(Separate neutral (N) and protective conductors (PE) within installation)

TN-C-S earthing system

OUTCOME 3

Understand the principles for selecting cables and circuit protection devices

CHOOSING WIRING SYSTEMS AND ENCLOSURES

Regulation 5 of the Electricity at Work Regulations 1989 states that 'No electrical equipment shall be put into use where its strength and capability may be exceeded in such a way as may give rise to danger.'

This is a very broad duty, as the term 'strength and capability' of electrical equipment refers to the ability of the equipment to withstand the thermal, electromagnetic, electro-chemical or other effects of the electrical currents that might be expected to flow when the equipment is part of a system.

The duty holder, and the system designer in particular, need to ensure that all system components selected are appropriate to the intended use and rated accordingly, bearing in mind that external influences have an impact on the wiring systems and enclosures chosen for any particular installation.

Assessment criteria

3.1 Explain how external influences can affect the choice of wiring systems and enclosures

SmartScreen Unit 304
Handout 6 and Worksheet 6

External influences

The wiring system must be able to withstand all the external influences it will be subjected to, including those listed in the table below. Particular care must be taken when a system enters a building or changes direction.

External influences on a wiring system

External influence	Example
Ambient temperature	The general constant temperature of the environment
External heat sources	Proximity to hot-water pipes, heaters
Water or high humidity	Immersed cables in a marina, condensation in an unheated building
Solid foreign bodies	Dust
Corrosive or polluting substances	Substances found in a tannery, battery room, plating plant, cowshed

ACTIVITY

Look up Section 522 of BS 7671 and name three factors that need to be considered when selecting a wiring system (main regulation headings).

KEY POINT

BS 7671 numbering system

In the numbering system used, the first digit signifies a Part, the second digit a Chapter, the third digit a Section and the subsequent digits the Regulation number. For example, the Section number 413 is made up as follows:

■ Part 4 – Protection for safety

■ Chapter 41 (first chapter of Part 4) – Protection against electric shock

■ Section 413 (third section of Chapter 41) – Protective measure: electrical separation.

External influence	Example
Impact	Vehicles in a garage or car park
Vibration	Connections to motors
Other mechanical stresses	Bends too tight to pull in cables, insufficient supports
Presence of flora or mould	Plants or moss
Presence of fauna	Vermin, livestock
Seismic effects	In locations susceptible to earthquakes
Movement of air	Sufficient to stress mountings
Nature of processed or stored materials	Risk of fire or degradation of insulation, for example, from oil spills, acid in plating shops
Building design	From building movement

ASSESSMENT GUIDANCE

All installations will use many different wiring and cable types.

Which wiring system to use?

General uses of the more common wiring systems are described in the table below.

Common wiring systems

Wiring system	Requirements	Common use
Clipped direct	Sheathed without armour	General application in domestic and commercial installations. Not usually acceptable to the client in new installations. Mechanical protection may be required in some locations by the installation of guards or the use of armoured cable
Clipped direct	Sheathed with armour	Commonly adopted in industrial premises where presence of cables is acceptable
Installed in the building structure	If using insulated and sheathed cables (that is, not armoured or enclosed in conduit, etc), precautions must be taken to prevent damage	Insulated and sheathed cables installed in metal-framed walls will probably require additional protection
Buried cables	Buried cables must be enclosed in a conduit or duct or be armoured. A sufficient depth is required to prevent damage by reasonably foreseeable disturbance to the ground	This is the most practical solution for large-sized cables for power distribution in a supply company network or between buildings
Plastic conduit systems	A very wide range of plastic conduits is available. Suitable for almost every location	Manufacturer's instructions must be read to ensure the conduit is suitable for the particular environment. Expansion needs to be allowed for

Wiring system	Requirements	Common use
Steel conduit systems	Common types are black enamel and galvanised. Stainless steel conduits may be specified for onerous locations	Galvanised is suitable for onerous environments and may be selected where the good mechanical properties are necessary
Cable trunking systems	Insulated cables may be used (without sheath)	Used where a number of small cables need to be run
Cable ducting systems	The most practical system for distributing large cables through a building or site	Normally formed by concrete for large cables or circular pipe for smaller cables
Cable ladder, cable tray, cable brackets	Cables need further mechanical protection by a sheath as this is not a containment system	Practical solution for distributing large cables around a building or site
Mineral-insulated	Particular skills are required to install these cables	They have exceptional fire-resistance properties and so are used for circuits requiring high integrity under fire conditions
On insulators	Widely adopted for electricity supply distribution in rural areas	Cheaper than laying cables in the ground, but may be considered unsightly and are susceptible to damage by high vehicles. Care necessary in some locations, eg caravan sites, riverbanks, farms, etc
Support wire	Low cost and may avoid considerable disturbance	Check that the height is sufficient for all vehicle movements, etc

Application of cable types

The table below provides guidance on the use of cables for particular cable systems.

Cable use for particular cable systems

Type of cable	Uses	Comments
Thermoplastic, thermosetting or rubber-insulated non-sheathed	In conduit, cable ducting or trunking	■ Intermediate support may be required on long, vertical runs ■ 70 °C maximum conductor temperature for normal wiring grades including thermosetting types ■ Cables run in PVC conduit shall not operate with a conductor temperature greater than 70 °C

Type of cable	Uses	Comments
Flat thermoplastic or thermosetting insulated and sheathed	■ General indoor use in dry or damp locations; may be embedded in plaster ■ On exterior surface walls, boundary walls, etc ■ Overhead wiring between buildings ■ Underground in conduits or pipes ■ In building voids or ducts formed in-situ	■ Additional protection may be necessary where exposed to mechanical stresses ■ Protection from direct sunlight may be necessary ■ Black sheath colour is better for cables in sunlight ■ Unsuitable for embedding directly in concrete ■ May need to use hard-drawn (HD) copper conductors for overhead wiring
Split-concentric thermoplastic insulated and sheathed	General	■ Additional protection may be necessary where exposed to mechanical stresses ■ Protection from direct sunlight may be necessary ■ Black sheath colour is better for cables in sunlight
Mineral-insulated	General	With overall PVC covering where exposed to the weather or risk of corrosion, or where installed underground or in concrete ducts
Thermoplastic or XLPE-insulated, armoured, thermoplastic sheathed	General	■ Additional protection may be necessary where exposed to mechanical stresses ■ Protection from direct sunlight may be necessary ■ Black sheath colour is better for cables in sunlight

Notes:

1 The use of cable covers or equivalent mechanical protection is desirable for all underground cables that might otherwise subsequently be disturbed. Route-marker tape should also be installed, buried just below ground level.

2 Cables with thermoplastic insulation or sheath should preferably be installed only when the ambient temperature is above 0 °C and has been for the preceding 24 hours. Where they are to be installed during a period of low temperature, precautions should be taken to avoid risk of mechanical damage during handling. A minimum ambient temperature of 5 °C is advised for some types of thermoplastic insulated and sheathed cables. Manufacturer's information must be followed.

3 Cables should be suitable for the maximum ambient temperature and should be protected.

4 Thermosetting cable types (to BS 7211 or BS 5467) can operate with a conductor temperature of 90 °C. This must be limited to 70 °C when drawn into a conduit, etc, with thermoplastic insulated conductors or connected to electrical equipment (Regulation 512.1.5 and Table 52.1), or when such cables are installed in plastic conduit or trunking.

5 For cables to BS 6004, BS 6007, BS 7211, BS 6346, BS 5467 and BS 6724, further guidance may be obtained from those standards. Additional advice is given in BS 7540-2:2005 Guide to use for cables with a rated voltage not exceeding 450/750 V for cables to BS 6004, BS 6007 and BS 7211.

6 Cables for overhead wiring between buildings must be able to support their self-weight and any imposed wind or ice/snow loading. A catenary support is usual but hard drawn copper types may be used.

CURRENT RATINGS OF CIRCUIT PROTECTION DEVICES

Protective devices may be one or a combination of:

- fuses
- circuit breakers (CBs)
- residual current devices (RCDs).

Device ratings

Protective devices are available in a wide range of nominal ratings (I_n). Remember, the nominal rating is the value of current that can flow through the device in normal operating conditions without deterioration of the lifespan of the device. A device having a nominal rating of 32 A will not disconnect at 33 A and requires much more current for rapid disconnection.

Although Appendix 3 of BS 7671 shows a wide range of nominal ratings, many others are available.

SELECTING OVERCURRENT PROTECTION DEVICES

Overcurrent protection

Each installation and every circuit within an installation must be protected against overcurrent by devices that will operate automatically to prevent injury to persons and livestock and damage to the installation, including the cables. The overcurrent devices must be of adequate **breaking capacity** and be so constructed that they will interrupt the supply without danger. Cables must be able to carry these overcurrents without damage. Overcurrents may be fault currents or overload currents.

Fault currents

Fault currents arise as a result of a fault in the cables or the equipment. There is a sudden increase in current, perhaps 10 or 20 times the cable rating, the current being limited by the **impedance** of the supply, the impedance of the cables, the impedance of the fault and the impedance of the return path. The current should be of short duration, as the overcurrent device should operate.

Overload currents

Overload currents do not arise as a result of a fault in the cable or equipment. They arise because the current has been increased by the addition of further load. Overload protection is only required if

Assessment criteria

3.2 State the current ratings for different circuit protection devices

ACTIVITY

Using the internet or catalogues, look at the different current ratings for protective devices that are available in addition to those given in Appendix 3 of BS 7671.

Assessment criteria

3.3 Specify and apply the procedure for selecting appropriate overcurrent protection devices

Breaking capacity

This is the maximum value of current that a protective device can safely interrupt.

KEY POINT

Chapter 43 of BS 7671 gives the requirements for overcurrent protection.

Impedance

Impedance (symbol Z) is a measure of the opposition that a piece of electrical equipment or cable makes to the flow of electric current when a voltage is applied. Impedance is a term used for alternating current circuits. The elements of impedance are:

- resistance
- inductance
- capacitance.

KEY POINT

Section 434 of BS 7671 provides the requirements for protection against fault currents.

KEY POINT

Section 433 of BS 7671 provides the requirements for protection against overload currents.

ACTIVITY

Compare the cost of:

a) a 30 A rewireable fuse

b) a 30 A cartridge fuse

c) a 32 A circuit breaker.

overloading is possible. It would not be required for a circuit supplying a fixed load, although fault protection would be required.

The load on a circuit supplying, for example, a 7.2 kW shower will not increase unless the shower is replaced, when the adequacy of the circuit must be checked against the new load criteria.

A distribution circuit supplying a number of buildings could be overloaded by additional machinery being installed in one of the buildings supplied.

Overload currents are likely to be of the order of 1.5 to 2 times the rating of the cable, whereas fault currents may be of the order of 10 to 20 times the rating.

Overloads of less than 1.2 to 1.6 times the device rating are unlikely to result in operation of the device. Regulation 433.1 of BS 7671 requires that every circuit be designed so that small overloads of long duration are unlikely to occur.

It is usual for one device in the circuit to provide both fault protection and overload protection. A common exception is the overcurrent devices to motor circuits, where the overcurrent device at the origin of the circuit provides protection against fault currents and the motor starter will be providing protection against overload.

Selecting protective devices

The type of protective device chosen will depend on a number of factors, including:

- the nature or type of load
- the prospective fault current I_{pf} at that point of the installation
- any existing equipment
- the user of the installation, as a CB is easier to reset than a bolted-type HRC fuse.

Type of load	Suitable type of device
Resistive, such as heating elements, incandescent lighting, etc	Type B circuit breakers gG BS 88 devices
Inductive, such as discharge lighting, small motors or ELV lighting transformers	Type C circuit breakers gM BS 88 devices
High inductive or surging, such as welding equipment, X-ray machines, large motors without soft starting	Type D circuit breakers

Protective devices are also selected for suitability against prospective fault current (PFC).

Breaking capacity

There is a limit to the maximum current that an overcurrent protective device (fuse or circuit breaker) can interrupt. This is called the rated short-circuit capacity or breaking capacity. The table on page 292 shows the rated short-circuit capacities of the most common devices used in the UK.

Regulation 434.1 in BS 7671 requires the prospective fault current under both short-circuit and earth-fault conditions to be determined at every relevant point of the complete installation. This means that at every point where switchgear is installed, the maximum fault current must be determined to ensure that the switchgear is adequately rated to interrupt the fault currents.

From the table on page 292 it can be seen that BS 3036 devices have a very low breaking capacity, compared to other devices. This is the main reason why these devices are no longer suitable for many installations.

Circuit breakers have two short-circuit capacity ratings.

- I_{cs} is the value of fault current up to which the device can operate safely and remain suitable and serviceable after the fault.

- I_{cn} is the value above which the device would not be able to interrupt faults safely. This could lead to the danger of explosion during faults of this magnitude or, even worse, the contacts welding and not interrupting the fault.

Any faults that occur between these two ratings will be interrupted safely but the device will probably require replacement.

ACTIVITY

BS 3871 circuit breakers and BS 1361 fuses are no longer listed in BS 7671. Why is it necessary to include them here?

Rated short-circuit capacities of protective devices (Source: IET On-Site Guide)

Device type	Device designation	Rated short-circuit capacity (kA)
Semi-enclosed fuse to BS 3036 with category of duty	S1A S2A S4A	1.0 2.0 4.0
Cartridge fuse to BS 1361 type I Cartridge fuse to BS 1361 type II Cartridge fuse to BS 88-3 type I Cartridge fuse to BS 88-3 type II		16.5 33.0 16.0 31.5
General purpose fuse to BS 88-6		50 at 415 V 16.5 at 240 V 80.0 at 415 V
Circuit breakers to BS 3871	M1 M1.5 M3 M4.5 M6 M9	1.0 1.5 3.0 4.5 6.0 9.0
Circuit breakers to BS EN 60898*		I_{cn} I_{cs} 1.5 (1.5) 3.0 (3.0) 4.5 (4.5) 6.0 (6.0) 10.0 (7.5) 15.0 (7.5) 20.0 (10.0) 25.0 (12.5)

Notes:

*Two short-circuit ratings are defined in BS EN 60898 and BS EN 61009

(a) I_{cn} is the rated short-circuit capacity (marked on the device)

(b) I_{cs} is the service short-circuit capacity

These are based on the condition of the circuit breaker after manufacturer's testing.

- I_{cn} is the maximum fault current the breaker can interrupt safely, although the breaker may no longer be usable.

- I_{cs} is the maximum fault current the breaker can interrupt safely without loss of performance.

The I_{cn} value is normally marked on the device in a rectangle, eg: 6000

For the majority of applications the prospective fault current at the terminals of the circuit breaker should not exceed this value.

For domestic installations the prospective fault current is unlikely to exceed 6 kA, up to which value the I_{cn} and I_{cs} values are the same.

DIVERSITY FACTOR AND ESTIMATING MAXIMUM DEMAND

Diversity factor is the ratio of the sum of the individual maximum electrical demands of all the circuits in an installation to the maximum demand of the whole system being considered. When measured, a diversity factor is usually more than one. The sum of the connected loads supplied by a circuit is normally multiplied by the demand factor to calculate the load used to size the components of the installation.

Installation outline

When the position and demand, in amperes (A) and kilovolt-amperes (kVA), of the installation have been ascertained (from drawings, specifications, discussion with the client, etc), an installation outline can be prepared. This will show the main and sub-distribution boards as well as the sub-mains that link the sub-distribution boards with the main board, as illustrated below.

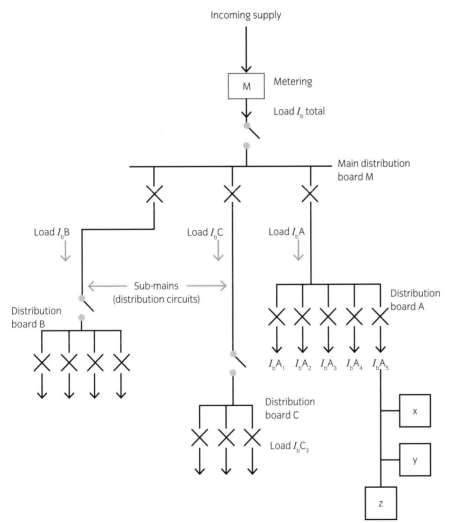

Installation outline

Assessment criteria

3.4 State what is meant by diversity factors and explain how a circuit's maximum demand is established after diversity factors are applied

SmartScreen Unit 304
Handout 14 and Worksheet 14

KEY POINT

BS 7671 Regulation 132.3: Nature of demand

The number and type of circuits required for lighting, heating, power, control, signalling, communication and information technology, etc, shall be determined from knowledge of:

- location of points of power demand
- loads to be expected on the various circuits
- daily and yearly variation of demand
- any special conditions such as harmonics
- requirements for control, signalling, communication and information technology, etc
- anticipated future demand if specified.

ACTIVITY

Calculating the size of each main and distribution cable takes a long time. Any changes may mean that you have to recalculate the whole lot. Cable design packages are readily available and will recalculate the whole installation in a few seconds.

Look on the internet for companies offering such packages.

For ease of drawing, this is normally a schematic. For an existing building, cable route lengths will be estimated by scaling off a map or even by direct measurement with a tape measure.

For a domestic installation the outline will be brief, as shown below.

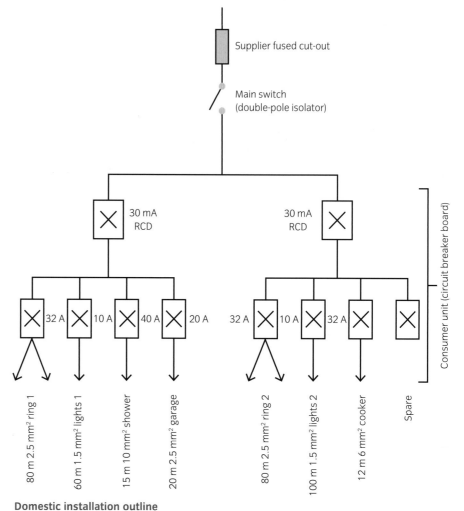

Domestic installation outline

Calculating current from kVA and kW

The current demand for an item of equipment for a given kVA rating can be calculated as follows.

For a single-phase supply the current I is given by:

$$I = \frac{kVA \times 1000}{U}$$

where:

I is current in amperes
kVA is the rating of the equipment
U is the nominal voltage of the equipment.

When the rating of the equipment is in kW, the current I is given by:

$$I = \frac{kW \times 1000}{U \times \cos \theta}$$

where:

I is current in amperes
kW is the rating of the equipment in kW
U is the nominal voltage of the equipment
$\cos \theta$ is the power factor.

For three-phase equipment, there are two methods for finding the current I.

Method 1:

$$I = \frac{kVA \times 1000}{3U}$$

where:

I is current in amperes
kVA is the rating of the equipment
U is the nominal voltage of the equipment.

Method 2:

$$I = \frac{kVA \times 1000}{\sqrt{3}U}$$

where:

I is current in amperes
kVA is the rating of the equipment
U is the nominal voltage (line-to-line) of the equipment.

Where the design current for discharge lighting is determined, a factor of 1.8 is applied if particular detail is not available (such as power factor ratings or particular starting current demands).

> **ASSESSMENT GUIDANCE**
>
> It is more common to use the line voltage when carrying out three-phase calculations.

Final circuit current demand

The current demand of a final circuit is estimated by adding the current demands of all points of utilisation (eg socket outlets) and items of equipment connected to the circuit and, where appropriate, making allowances for diversity.

For domestic and similar small installations, the table below gives current demands to be used for final circuits. The rated current given in item 1 relates to the nominal rating of the protective device selected (I_n).

Final circuit current demand to be assumed for points of utilisation and current-using equipment (IET On-Site Guide, Table A1)

	Point of utilisation or current-using equipment	Current demand to be assumed
1	Socket outlets other than 2 A socket outlets and 13 A socket outlets (see Note 1 below)	Rated current
2	2 A socket outlets	At least 0.5 A
3	Lighting outlet (see Note 2 below)	Current equivalent to the connected load, with a minimum of 100 W per lampholder
4	Electric clock, shaver supply unit (complying with BS 3535), shaver socket outlet (complying with BS 4573), bell transformer and current-using equipment of a rating not greater than 5 VA	May be neglected
5	Household cooking appliance	The first 10 A of the rated current plus 30% of the remainder of the rated current plus 5 A if a socket outlet is incorporated in the control unit
6	All other stationary equipment	Normal current or British Standard rated current

Note 1: See Appendix H (of the IET On-Site Guide) for the design of standard circuits using socket outlets to BS 1363-2 and BS 4343.

Note 2: Final circuits for discharge lighting must be arranged so as to be capable of carrying the total steady current, namely that of the lamp(s) and any associated gear and also their harmonic currents. Where more exact information is not available, the demand, in volt-amperes (VA), is taken as the rated lamp watts multiplied by not less than 1.8. This multiplier is based upon the assumption that the circuit is corrected to a power factor of not less than 0.85 lagging, and takes into account control gear losses and harmonic current.

ACTIVITY

2 A sockets were often used to connect table lamps or similar, the socket being controlled from a wall switch. The current equivalent was based on a 100 W lamp. Tungsten lamps of that rating are no longer available, and have typically been replaced by 18 W fluorescent lamps. What would be the current demand at 230 V of:

a) a 100 W tungsten lamp

b) a 18 W compact fluorescent?

KEY POINT

Table A1 in the IET On-Site Guide may be applied to final circuits but Table A2 is only applied to a collection of final circuits in order to size a distribution circuit or overall maximum demand including diversity.

Examples of circuit current demand

Shower circuit

(See row 6 of the table on page 300 or Table A1 of the IET On-Site Guide.)

Rating of shower (P) from technical specification 7.9 kW at 230 V:

$$\text{Current demand } I = \frac{kW \times 1000}{U \times \cos\theta}$$

$\cos\theta = 1$ for a resistive load

$$\text{current demand } I = \frac{7.9 \times 1000}{230} = 34.3 \text{ A}$$

> **KEY POINT**
>
> Certain loads have no diversity because they are either 100% off or 100% on.

Cooker circuit

(See row 5 of the table on page 300 or Table A1 of the IET On-Site Guide.)

Consider an electric cooker with:

- hob comprising 4 of 2 kW elements
- main oven 2 kW
- grill/top oven 2 kW
- total installed capacity of 12 kW at 230 V.

$$I = \frac{kW \times 1000}{U \times \cos\theta} = \frac{12 \times 1000}{230 \times 1}$$

So the short-time peak demand = 52.2 A

From row 3 of the table on page 300, the circuit design current (I_b) applying diversity is the first 10 A of the rated current plus 30% of the remainder of the rated current plus 5 A if a socket outlet is incorporated in the control unit.

$$\text{After diversity demand } I = 10 + \left(\frac{30}{100}\right) \times (52.2 - 10) = 22.7 \text{ A}$$

Or as an alternative method:

$$I_b = 52.2 - 10 = 42.2 \times \left(\frac{30}{100}\right) + 10 = 22.7 \text{ A}$$

After diversity is applied:

$$I_b = 22.7 \text{ A (or +5 A if a socket outlet is included = 27.7 A)}$$

Lighting circuit

(See row 3 of the table on page 300 or Table A1 of the IET On-Site Guide.)

Consider 10 downstairs lights with standard brass bayonet cap (BC) lampholders. In line with the table of final circuit current demand to be assumed for points of utilisation and current-using equipment on page 296, assume 100 W demand per lighting point. Therefore:

$$\text{Circuit demand } I_b = \frac{\text{total load in watts}}{\text{voltage}} = \frac{10 \times 100}{230} = 4.3 \text{ A}$$

A 6 A circuit would be suitable for tungsten lamps. If extra low voltage or discharge lighting is to be supplied by a type B circuit breaker, you would need to specify a 10 A circuit to reduce unwanted tripping on switch on, or a 6 A type C circuit breaker due to starting surge.

For all but the simplest circuits the load characteristics should be assessed and the manufacturer's data applied for the selection of circuit breakers. Manufacturer's data should be consulted in particular for tungsten flood lamps, heat lamps, discharge lighting or transformers.

If detailed information is not available for circuits containing discharge luminaires, the rating of the luminaires should be multiplied by 1.8 to determine the circuit design current.

Immersion heater circuit
(See row 6 of the table on page 300.)

Consider a single-phase 230 V, 3 kW immersion heater circuit:

$$I = \frac{3 \times 1000}{230 \times 1} = 13.04 \text{ A}$$

Motor circuit
Consider a single-phase 230 V, 3 kW motor with power factor $\cos \theta = 0.8$:

$$I_b = \frac{\text{kW} \times 1000}{U \times \cos \theta} = \frac{3 \times 1000}{230 \times 0.8} = 16.3 \text{ A}$$

For three-phase equipment, the design current I_b is given by:

$$I_b = \frac{\text{kVA} \times 1000}{3U}$$

Where:
I_b is design current (amperes)
kVA is the rating of the equipment
U is the nominal voltage of the equipment.

Or more commonly:

$$I_b = \frac{\text{kVA} \times 1000}{\sqrt{3}U}$$

Where:

I_b is design current (amperes)
kVA is the rating of the equipment
U is the nominal voltage (line to line) of the equipment.

The above calculation is for balanced three-phase loads. The design current determined is per line; in other words, the current that each cable is sized for.

Where three-phase loads are unbalanced, they should be treated as three individual single-phase loads and the circuit cables sized on the largest of the three loads determined.

Estimating the demand of a distribution circuit supplying a number of final circuits

The current demand of a distribution circuit that supplies a number of final circuits may be assessed by using the allowances for diversity given in the table of allowances for diversity between final circuits for sizing distribution circuits on page 300 (extract from the IET On-Site Guide, Table A2); you apply these allowances to the total current demand of all the equipment supplied by that circuit.

You do not assess the current demand of the distribution circuit by adding the current demands of the individual final circuits (as outlined above). In the table shown on page 300 (extract from the IET On-Site Guide, Table A2), the allowances are expressed either as percentages of the current demand or – where followed by the letters f.l. (full load) – as percentages of the rated full-load current of the current-using equipment.

The current demand for any final circuit that is a conventional circuit arrangement, and which complies with Appendix H of the IET On-Site Guide, is the rated current of the overcurrent protective device of that circuit. For example, a 20 A radial circuit that supplies socket outlets would be protected by a 20 A circuit breaker, so the design current would be assessed as being 20 A.

An alternative method of assessing the current demand of a circuit that supplies a number of final circuits is to add the diversified current demands of the individual circuits and then apply a further allowance for diversity, in the form of a factor. In this method, the allowances given in Table A2 of the IET On-Site Guide are not to be used; rather, the values to be chosen are the responsibility of the installation designer. This may be particularly useful when assessing the total demand for the supply to a building that may contain many circuits, but consumption is likely to be low. This method is only recommended for experienced designers.

As the value of supply voltage has an effect on the design current, some situations may require the designer to use actual values of voltage relevant to the installation instead of the standard 230 V.

ACTIVITY

Using BS 7671 and the On-Site Guide, fill in the blanks in the table as best you can. Use Table 4D5, etc and Method C.

I_b load	I_n BS EN 60898	I_t cable pvc/pvc
15		
25	32	
3	6	
60		
41		

Allowances for diversity between final circuits for sizing distribution circuits (IET On-Site Guide, Table A2)

	Purpose of final circuit fed from conductors or switchgear to which diversity applies	**Type of premises** Individual household installations including individual dwellings of a block	**Type of premises** Small shops, stores, offices and business premises	**Type of premises** Small hotels, boarding houses, guest houses, etc
1	Lighting+	66% of total current demand	90% of total current demand	7% of total current demand
2	Heating and power (but see 3 to 8 below)+	100% of total current demand + up to 10 A +50% of any current demand in excess of 10 A	100% full load (f.l.) of largest appliance +75% f.l. of remaining appliances	100% f.l. of largest appliance +80% f.l. of second largest appliance +60% f.l. of remaining appliances
3	Cooking appliances	10 A +30% f.l. of connected cooking appliances in excess of 10 A +5 A if socket outlet incorporated in control unit	100% f.l. of largest appliance +80% f.l. of second largest appliance +60% f.l. of remaining appliances	100% f.l. of largest appliance +80% f.l. of second largest appliance +60% f.l. of remaining appliances
4	Motors (other than lift motors that are subject to special consideration)	Not applicable	100% f.l. of largest motor +80% f.l. of second largest motor +60% f.l. of remaining motors	100% f.l. of largest motor +50% f.l. of remaining motors
5	Water-heaters (instantaneous type)*	100% f.l. of largest appliance +100% f.l. of second largest appliance +25% f.l. of remaining appliances	100% f.l. of largest appliance +100% f.l. of second largest appliance +25% f.l. of remaining appliances	100% f.l. of largest appliance +100% f.l. of second largest appliance +25% f.l. of remaining appliances
6	Water-heaters (thermostatically controlled)	No diversity allowable†	No diversity allowable†	No diversity allowable†
7	Floor warming installations	No diversity allowable†	No diversity allowable†	No diversity allowable†
8	Thermal storage space heating installations	No diversity allowable†	No diversity allowable†	No diversity allowable†
9	Standard arrangement of household and similar final circuits (in accordance with Appendix 8 of the IET On-Site Guide)+	100% of current demand of largest circuit +40% of current demand of every other circuit	100% of current demand of largest circuit +50% of current demand of every other circuit	100% of current demand of largest circuit +50% of current demand of every other circuit

	Purpose of final circuit fed from conductors or switchgear to which diversity applies	**Type of premises** Individual household installations including individual dwellings of a block	**Type of premises** Small shops, stores, offices and business premises	**Type of premises** Small hotels, boarding houses, guest houses, etc
10	Socket outlets other than those included in 9 above and stationary equipment other than those listed above	100% of current demand of largest point of utilisation +40% of current demand of every other point of utilisation	100% of current demand of largest point of utilisation +70% of current demand of every other point of utilisation	100% of current demand of largest point of utilisation +75% of current demand of every other point in main rooms (dining rooms, etc) +40% of current demand of every other point of utilisation

* For the purpose of this table, an instantaneous water-heater is deemed to be a water-heater of any loading that heats water only while the tap is turned on and therefore uses electricity intermittently.

† It is important to ensure that the distribution boards and consumer units are of sufficient rating to take the total load connected to them without the application of any diversity.

+ The current demand may be that estimated, for example in accordance with Table A1 IET On-Site Guide. Where the circuit is a standard circuit for household or similar installations, the current demand is the rated current of the overcurrent protective device of the circuit.

SELECTING CABLE SIZE

In this section we will look at various methods of determining a suitable conductor cross-sectional area (csa). These will include:

Assessment criteria

3.5 Specify and apply the procedure for selecting a suitably sized cable

- a simple single circuit method using the IET On-Site Guide including voltage drop
- detailed methods using BS 7671 where overload and short-circuit protection is required
- detailed methods using BS 7671 where only short-circuit protection is provided
- verification of voltage drop using BS 7671.

Selecting cable size using the IET On-Site Guide

The following procedures use information from the IET On-Site Guide as it provides a simplistic approach for simple single circuit design. More complex circuit design procedures are covered in Unit 305 (2365 Level 3) which involves detail from BS 7671.

Before looking at the procedure for selecting the correct conductor csa, we will consider the factors that affect the ability of a conductor to carry current.

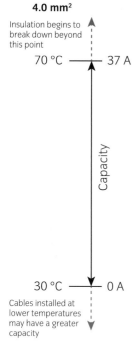

4.0 mm²

Insulation begins to break down beyond this point

70 °C ———— 37 A

Capacity

30 °C ———— 0 A

Cables installed at lower temperatures may have a greater capacity

PVC/PVC flat twin, 70 °C thermoplastic, 30 °C ambient

Conductor insulation

It is the conductor insulation that governs the overall capacity of a conductor, as it is the weak point. Copper can carry large currents before it reaches melting point but the insulation cannot. General thermoplastic insulation has a maximum operating temperature of 70 °C so it is this temperature that sets the current limit. It is assumed that the initial temperature of any conductor, and therefore insulation, is 30 °C before any current is put through the conductor. As soon as current flows, the conductor temperature increases. The amount of current causing the temperature to reach 70 °C is the upper limit of the conductor.

Looking at Table F6 of the IET On-Site Guide below, it can be seen that a single-phase 4 mm² flat-profile cable installed in free air (method C, column 6) has a tabulated capacity of 37 A. This essentially means that, given no further influences, 37A would cause the conductor's temperature to rise from 30 °C to 70 °C.

▼ **Table F6**   70 °C thermoplastic (PVC) insulated and sheathed flat cable with protective conductor (copper conductors)

Ambient temperature: 30 °C
Conductor operating temperature: 70 °C

Current-carrying capacity (amperes) and voltage drop (per ampere per metre):

Conductor cross-sectional area	Reference method 100* (above a plasterboard ceiling covered by thermal insulation not exceeding 100 mm in thickness)	Reference method 101* (above a plasterboard ceiling covered by thermal insulation exceeding 100 mm in thickness)	Reference method 102* (in a stud wall with thermal insulation with cable touching the inner wall surface)	Reference method 103 (in a stud wall with thermal insulation with cable not touching the inner wall surface)	Reference method C (clipped direct)	Reference method A (enclosed in conduit in an insulated wall)	Voltage drop
1	2	3	4	5	6	7	8
mm²	A	A	A	A	A	A	mV/A/m
1	13	10.5	13	8	16	11.5	44
1.5	16	13	16	10	20	14.5	29
2.5	21	17	21	13.5	27	20	18
4	27	22	27	17.5	37	26	11
6	34	27	35	23.5	47	32	7.3
10	45	36	47	32	64	44	4.4
16	57	46	63	42.5	85	57	2.8

Notes:
*   Reference methods 100, 101 and 102 require the cable to be in contact with the plasterboard ceiling, wall or joist, see Tables 7.1(ii) and 7.1(iii) in Section 7.
1   Wherever practicable, a cable is to be fixed in a position such that it will not be covered with thermal insulation.
2   Regulation 523.9, BS 5803-5: Appendix C: Avoidance of overheating of electric cables, Building Regulations Approved Document B and Thermal insulation: avoiding risks, BR 262, BRE, 2001 refer.

Current-carrying capacity (from Table F6 of the IET On-Site Guide)

Ambient temperature

If cables are installed in ambient temperature other that 30 °C, the capacity of the cable's conductor is altered. If a cable was installed in an ambient temperature higher than 30 °C, the capacity of the conductor will be reduced as the temperature of the cable, before any current flows, is higher. Rating factors (C_a) are used from Table F1 of the IET On-Site Guide. These rating factors are used to alter the capacity of the conductor depending on the ambient temperature of the cable and the type of insulation applied to the conductor. For any cable run which passes through a range of ambient temperatures, the highest shall be used to obtain a rating factor.

▼ **Table F1**   Rating factors (C_a) for ambient air temperatures other than 30 °C to be applied to the current-carrying capacities for cables in free air

Ambient temperature (°C)	70 °C thermoplastic	90 °C thermosetting	Mineral	
			Thermoplastic covered or bare and exposed to touch 70 °C	Bare and not exposed to touch 105 °C
25	1.03	1.02	1.07	1.04
30	1.00	1.00	1.00	1.00
35	0.94	0.96	0.93	0.96
40	0.87	0.91	0.85	0.92

Ambient temperatures (from Table F1 of the IET On-Site Guide)

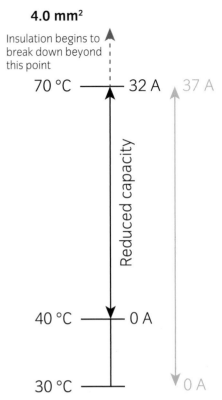

4.0 mm²

Insulation begins to break down beyond this point

70 °C — 32 A   37 A

Reduced capacity

40 °C — 0 A

30 °C — 0 A

PVC/PVC flat twin, 70 °C thermoplastic, 40 °C ambient

Thermal insulation

Where a cable is totally surrounded by thermal insulation, the cable's ability to dissipate heat is greatly reduced for any conductor that has a csa of 10 mm² or less. As a result, rating factors for thermal insulation (C_i) need to be applied as the heat build-up in the cable will have an effect on the temperature of the cable's insulation. Ideally, cables should be clipped to a joist, or other thermally conducting surfaces, or should be sleeved by conduit as they pass through the thermal insulation. In these situations, a thermal insulation rating factor need not be applied as the current capacity is allowed for within the selection tables depending on the method of insulation. As can be seen from Table F2 of the IET On-Site Guide below, a rating factor of 0.5 is applied to any cable which passes through 500 mm or more thermal insulation. This effectively reduces a conductor's capacity by 50 % so thermal insulation is best avoided if possible.

▼ **Table F2**   Cable surrounded by thermal insulation

Length in insulation (mm)	Derating factor (C_i)
50	0.88
100	0.78
200	0.63
400	0.51
≥ 500	0.50

Thermal insulation (from Table F2 of the IET On-Site Guide)

Grouping of circuits

Where circuits are grouped together, either in the same containment system such as conduit, or clipped close together on a wall, the heat from one circuit can have an effect on others. Rating factors for grouping (C_g) are applied to circuits using Table F3 of the IET On-Site Guide (see below). As can be seen from the table, the more circuits that are grouped, the more the conductor's capacity is reduced. As an example, if a new circuit was installed into a cable basket with six other circuits, that would be seven in total. From the table, cables bunched in air (methods A to F) with a total number of circuits being seven has a rating factor of 0.54. This means that the conductor's capacity is reduced to 54 % of the tabulated value.

Where the horizontal clearance between cables exceeds twice their diameter no factor need be applied. Cables are assumed to be equally loaded. Where cables are expected to carry no more than 30% of their grouped rating they may be ignored so far as grouping is concerned. The IET On-Site Guide contains more conditions regarding grouping.

▼ **Table F3** Rating factors (C_g) for one circuit or one multi-core cable or for a group of circuits, or a group of multi-core cables (to be used with the current-carrying capacities of Tables F4(i), F5(i) and F6)

Arrangement (cables touching)	Number of circuits or multi-core cables										Applicable reference method for current-carrying capacities
	1	2	3	4	5	6	7	8	9	12	
Bunched in air, on a surface, embedded or enclosed	1.0	0.80	0.70	0.65	0.60	0.57	0.54	0.52	0.50	0.45	A to F
Single layer on wall or floor	1.0	0.85	0.79	0.75	0.73	0.72	0.72	0.71	0.70	0.70	C
Single layer multicore on a perforated horizontal or vertical cable tray system	1.0	0.88	0.82	0.77	0.75	0.73	0.73	0.72	0.72	0.72	E
Single layer multicore on a cable ladder system or cleats, etc.	1.0	0.87	0.82	0.80	0.80	0.79	0.79	0.78	0.78	0.78	E

Grouping of circuits (from Table F3 of the IET On-Site Guide)

Method of installation

Basic methods of installation are listed below but for a concise list, see Table 4A2 in Appendix 4 of BS 7671. The method used to install cables can affect the ability of a cable to dissipate heat, meaning that cables could run too hot if the correct method or table selection isn't considered. When selecting a conductor csa from the relevant current capacity table, each column represents a different method of installation. The common methods include:

- **Method A**: enclosed in conduit in an insulated wall.
- **Method B**: enclosed in surface-mounted conduit.
- **Method C**: clipped directly to a surface in free air.

- **Method 100**: in contact with wooden joists or a plaster ceiling and covered by thermal insulation *not exceeding* 100 mm thickness.
- **Method 101**: in contact with wooden joists or a plaster ceiling and covered by thermal insulation *exceeding* 100 mm thickness.
- **Method 102**: in a stud wall containing thermal insulation with the cable touching the wall.
- **Method 103**: in a stud wall containing thermal insulation with the cable *not* in contact with the wall.

In order to determine a suitable conductor csa for a single circuit cable, the following procedure is used.

Step 1

Determine the design current (I_b). This is determined in several ways depending on the type of load and the allowance, if any, for diversity.

> **KEY POINT**
>
> Diversity is an allowance to reduce (or sometimes increase) design current for a final or distribution circuit, based on the assumption that not all loads on that circuit will be operating at the same time, based on the type of installation. The table below shows how to determine design current for simple single-phase final circuits.

Load type and nature of the final circuit	How to determine design current (I_b)	Notes
Standard resistive loads such as water heaters, electric space heaters and incandescent lighting	$I_b = \dfrac{\text{total watts}}{\text{volts}}$	For ceiling roses and similar in dwellings, each point is assumed as 100 W. See Table A1, IET On-Site Guide
Socket-outlet circuits such as radial or ring-final circuits	$I_b = I_n$	Where I_n is the rating of protective device
For discharge lighting such as circuits containing many fluorescent luminaires	$I_b = \dfrac{\text{total watts} \times 1.8}{230}$	The factor of 1.8 allows for increased current during starting which all cables and switching devices must be capable of handling
For inductive loads where power factor is known, such as motors	$I_b = \dfrac{\text{total watts}}{\text{volts} \times \cos\theta}$	Where $\cos\theta$ is the power factor
Domestic cooker circuits	$I_b = \left(\dfrac{\text{total watts}}{\text{volts}} - 10 \right) \times 0.3 + 10$	See Table A1, IET On-Site Guide Add 5 A if the cooker control incorporates a socket outlet
Other	See Table A1, IET On-Site Guide	

How to determine design current for simple single-phase final circuits

Step 2

Select the nominal rating of protective device (I_n).

$$I_n \geq I_b$$

Step 3

Gather information such as:

- Circuit length: this is the full length of the circuit from the distribution board to the final point of use on the circuit in metres.

- Method of installation (see Section 7, IET On-Site Guide).

- Rating factor for ambient temperature C_a (Table F1, IET On-Site Guide).

- Rating factor for thermal insulation C_i (Table F2, IET On-Site Guide).

- Rating factor for grouping of circuits C_g (Table F3, IET On-Site Guide).

- Rating factor to be applied if the protective device is a BS 3036 semi-enclosed fuse (C_f) = 0.725 (see page 267).

> **KEY POINT**
>
> If a cable is not grouped or surrounded by thermal insulation, a factor of 1 is used. In reality, this can be ignored from any calculations as multiplying or dividing any value by 1 remains unchanged.

Step 4

Apply the following calculation:

$$I_t \geq \frac{I_n}{C_a \times C_i \times C_g \times C_f}$$

Where I_t is the tabulated current-carrying capacity specified (see step 5).

Note: any factor which is 1 may be ignored.

Step 5

Use the correct IET On-Site Guide cable selection table from:

- Table F4(i) for single-core cables having 70 °C thermoplastic insulation

- Table F5(i) for multi-core cables (non-armoured) having 70 °C thermoplastic insulation

- Table F6 for multi-core flat profile cables having 70 °C thermoplastic insulation.

Select, using the correct column number relating to the installation method, a value of tabulated current equal to or larger than that calculated in step 4.

This will now ensure that the cable selected is of suitable current-carrying capacity.

Step 6

Taking voltage drop into account, ensure that the selected csa is appropriate. For the cable csa selection tables used in step 5, each has a corresponding voltage drop table or column. The values given in these tables or columns represent the value of voltage drop, in milli-volts per ampere per metre of circuit (mV/A/m). The circuit length determined in step 3 is required to determine voltage drop but remember that the circuit (cable) length is the length of the line conductor with the neutral, not added to the neutral, eg if a twin cable has a length of 10 m, that is the length of the line with neutral. The line added to the neutral would be 20 m (there and back). The values of voltage drop per metre include the neutral resistance within the values.

Apply the following formula to determine the value of voltage drop:

$$\text{voltage drop} = \frac{I_b \times \text{length} \times \text{mV/A/m}}{1000} \text{ (volts)}$$

Currently, BS 7671 states that values of voltage drop for an installation supplied from a public network should not exceed:

- 3% of the supply voltage for lighting circuits (3% of 230 V = 6.9 V)
- 5% of the supply voltage for power circuits (5% of 230 V = 11.5 V).

As long as the voltage drop calculated for the selected cable csa is within the values above, the cable selected is suitable. If the calculated voltage drop is too high, the next largest csa is selected and calculated for voltage drop until a suitable csa is found.

Note: if a final circuit is supplied by a distribution circuit, the voltage drop for the distribution circuit will also need to be taken into account.

Selecting cable size using BS 7671

We have seen how to determine the design current for a circuit on pages 295–299 and for distribution circuits on pages 299–301 of this book. For circuits containing socket outlets, the design current may be assumed to be the same as the protective device rating. For example, if a circuit is supplying a 16 A socket outlet, the device rating is likely to be 16 A so the design current may be assumed to be 16 A.

To determine the nominal rating of protective device I_n

Once the design current has been determined (see pages 295–296), the device rating can be selected. The device rating (or current setting) I_n must be greater than or equal to the design current:

$$I_n \geq I_b$$

For example, if the design current I_b is 29 A and a type B device is to be selected, I_n must be 32A.

As protective device nominal ratings (I_n) are the values that the device is rated for in normal service, then I_b can equal I_n.

Cable installation method

Like all equipment, the cabling system adopted for a location must be suitable for the environment. Table 4A2 of Appendix 4 of BS 7671 lists 120 cable installation methods.

The cable installation method needs to be known, as certain rating factors and cable selection tables depend on the type of method used.

Ambient temperature rating factor C_a

C_a from 4B1 of BS 7671 is based on ambient air temperatures around the wiring system. The worst-case temperature must be used. For example, if the cable passes through an airing cupboard in a house, the ambient temperature is likely to be 35 °C and that must be applied. The factor selected is also based on the type of cable insulation selected.

Ambient ground temperatures for buried cables are found in Table 4B2 of BS 7671.

Grouping rating factor C_g

Where more than one circuit is contained in a wiring system, or where cables on a tray or wall are close enough together to be classed as touching, rating factors for grouping must be applied. In order to reduce complicated calculations, circuits of similar size and type should be grouped together to keep them to a minimum. If cables of different type and size are enclosed together, further calculations, based on single or poly-phase circuits and additional factors, need to be applied. This is beyond the scope of this course.

Rating factors for grouping are based on the *total* number of circuits as well as the method of installation. Where cables are, for example, clipped to a wall, they are classed as spaced if the distance between them is more than twice the diameter of the cable. If not, they must be classed as grouped.

Thermal insulation factor C_i

C_i is a factor for thermal insulation.

For a single cable surrounded by thermally insulating material over a length of more than 0.5 m, the current-carrying capacity should be taken, in the absence of more precise information, as 0.5 times the current-carrying capacity of that cable clipped directly to a surface and open (Reference Method C). This means that cables in thermal insulation for this length will only be able to carry half the current of one clipped to a wall, so a cable will theoretically need to be twice as big.

Where a cable is to be totally surrounded by thermal insulation for less than 0.5 m, the current-carrying capacity of the cable should be reduced appropriately, depending on the size of cable, length in insulation and thermal properties of the insulation. The rating factors in Table 52.2 of BS 7671 are appropriate to conductor sizes up to 10 mm^2, installed and surrounded by thermal insulation having a thermal conductivity (λ) greater than 0.0625 Wm^{-1}K^{-1}. (Manufacturer's data relating to the thermal insulation needs to be checked.)

KEY POINT

Cables having cross-sectional area exceeding 10 mm^2 are large enough to conduct heat to a cooler environment, where the distance covered by thermal insulation is no greater than 0.5 m.

Length in insulation (mm)	Derating factor
50	0.89
100	0.81
200	0.68
400	0.55

Cable surrounded by thermal insulation (extract from Table 52.2 of BS 7671:2008 Requirements for Electrical Installations (the IET Wiring Regulations 17th edition) (2011))

If a cable is installed in a thermally insulated wall or ceiling, in line with one of the recognised methods indicated in the cable selection tables, the rating factors for thermal insulation, as shown on the previous page, do not need to be applied. For example:

■ Method 100: flat profile cable above a ceiling covered on one side only by thermal insulation not exceeding 100 mm in depth

■ Method 101: flat profile cable above a ceiling covered on one side only by thermal insulation exceeding 100 mm in depth

■ Method 102: flat profile cable in a thermally insulated stud wall with one side touching the inner wall surface

■ Method 103: cable in a thermally insulated stud wall not touching the inner wall surface

■ Method A: cable in conduit in an insulated wall.

The effect of the thermal insulation, as detailed above, is taken into consideration by the cable selection tables.

ASSESSMENT GUIDANCE

Method 103 covers cable in a studding wall surrounded by insulation (such as fibre glass or rockwool, etc) not touching the inner wall surface. Studding walls are normally constructed of timber or metal with plasterboard covering.

In situations not given above, it is often better to avoid thermal insulation or, if it is unavoidable, to sleeve the cable as it passes through any insulation, thus providing an air gap. This is only possible where protection against the spread of fire is not required.

Overcurrent device fuse rating factor C_f

Points i) and ii) of Regulation 433.1.1 in BS 7671 require that, where overload protection is necessary for any circuit:

$$I_n \geqslant I_b \qquad\qquad I_z \geqslant I_n$$

This is something we have already seen. Point iii) of Regulation 433.1.1 also states:

$$I_2 \leqslant 1.45 \times I_z$$

This means that the cable rating after factors have been applied, (I_z), must be capable of taking 45% more current, thereby ensuring effective disconnection of the protective device under overload (I_2) to guarantee the cable is suitable for any overload.

Equally, by transposing the formula, we can determine that the cable should be de-rated by:

$$\frac{1.45}{I_2}$$

> **KEY POINT**
>
> This is only required where the fusing factor of the overload device selected is greater than 1.45, for example, BS 3036 devices.

> **KEY POINT**
>
> Regulation 432.1 of BS 7671 gives a requirement for protection against overload current and fault current.

> **ASSESSMENT GUIDANCE**
>
> Any exam questions will only involve the four main correction factors: ambient, fuse, grouping and thermal insulation factors.

Protection against overload and short circuit

For circuits where overload and short circuit are likely, the following method is used to select the correct cross-sectional area (csa) of conductor.

When the device rating is known, the tabulated cable rating can be calculated. Where the overcurrent device (fuse or circuit breaker) provides protection against overload and short circuit, as is usual, then the tabulated current rating I_t:

$$I_t \geqslant \frac{I_n}{C_g C_a C_s C_d C_i C_f C_c}$$

where:

I_t is the tabulated current-carrying capacity of a cable, found in Appendix F of the IET On-Site Guide and Appendix 4 of BS 7671

I_n is the rated current or setting of a protective device

C_g is the rating factor for the grouping (see Table 4C1 of BS 7671 or F3 of the IET On-Site Guide)

C_a is the rating factor for the ambient temperature (see Table 4B of BS 7671 or F1 of the IET On-Site Guide)

C_s is the rating factor for the thermal resistivity of the soil

C_d is the rating factor for the depth of burial of a cable in the soil

C_i is the rating factor for conductors surrounded by thermal insulation

C_f is a rating factor applied when overload protection is being provided by overcurrent devices with fusing factors greater than 1.45 (eg C_f = 0.725 for semi-enclosed fuses to BS 3036)

C_c is the rating factor for circuits buried in the ground.

For cables not laid in the ground this simplifies to:

$$I_t \geqslant \frac{I}{C_g C_a C_i C_f}$$

ASSESSMENT GUIDANCE

A short circuit is a fault of negligible impedance between live conductors of the same circuit.

Example:

A circuit is to be installed in a surface-mounted trunking, which contains four other circuits. The circuit is to be wired using single-core 70 °C thermoplastic-insulated cable with copper conductors. The design current for the single-phase circuit (I_b) is 17 A, and the ambient temperature is 25 °C. Protection against overload and short circuit is by a Type B circuit breaker to BS EN 60898.

As $I_n \geqslant I_b$ then a 20 A device is selected.

Using the information contained in Appendix 4 of BS 7671, the method of installation is B from Table 4A2.

C_a= 1.03 from Table 4B1 for 25 °C and thermoplastic cable
C_g= 0.6 from Table 4C1, given the number of circuits is 5 in total, installed as method B
C_i= 1 as no thermal insulation exists
C_f= 1 as the device is *not* to BS 3036.

$$I_t \geqslant \frac{I}{C_g C_a C_i C_f}$$

Therefore:

$$I_t \geqslant \frac{20}{1.03 \times 0.6 \times 1 \times 1} = 32.36 \text{ A}$$

So, using the correct cable selection table for single-core 70 °C thermoplastic cable from Appendix 4 of BS 7671, we can select a conductor cross-sectional area (csa). Using Table 4D1A and, in particular, column 4, as the circuit is single-phase and the installation method is B, the value of I_t suitable for the current calculated is 41 A, which indicates a 6 mm^2 conductor size.

KEY POINT

Remember to use brackets on your calculator when you input the bottom line. For example, if brackets are not used on the bottom line of a calculation, the formula:

$\frac{20}{0.89 \times 0.9}$ becomes:

$\frac{20}{0.89} \times 0.9 = 20.22$ and is incorrect.

$\frac{20}{(0.89 \times 0.9)} = 24.96$, which is the correct answer

Also remember that any number multiplied or divided by 1 remains unchanged, so any value that is 1 can be disregarded in the calculation.

ACTIVITY

Determine the csa for a flat 70 °C thermoplastic cable supplying a 12 kW 230 V cooker circuit where the cooker control incorporates a 13 A socket outlet. The single circuit is to be wired clipped directly to a surface over a distance of 20 m.

Ambient temperature is 25 °C. Protection is by a circuit breaker to BS EN 60898.

Protection against short circuit only

Where the overcurrent device (fuse or circuit breaker) provides protection against short circuit only, as the load is fixed, the following formula is used:

$$I_t \geqslant \frac{I_b}{C_g C_a C_s C_d C_i C_f C_c}$$

where:

I_t is the tabulated current-carrying capacity of a cable, found in Appendix F of the IET On-Site Guide and Appendix 4 of BS 7671

I_b is the design current of the circuit

C_g is the rating factor for grouping (see Table 4C1 of BS 7671 or F3 of the IET On-Site Guide)

C_a is the rating factor for ambient temperature (see Table 4B of BS 7671 or F1 of the IET On-Site Guide)

C_s is the rating factor for the thermal resistivity of the soil

C_d is the rating factor for the depth of burial of a cable in the soil

C_i is the rating factor for conductors surrounded by thermal insulation

C_f is a rating factor applied when overload protection is being provided by overcurrent devices with fusing factors greater than 1.45 (eg C_f = 0.725 for semi-enclosed fuses to BS 3036)

C_c is the rating factor for circuits buried in the ground.

$$I_t \geqslant \frac{I_b}{C_g C_a C_i C_f}$$

Example:

A circuit is to be installed in a surface-mounted trunking, which contains four other circuits. The circuit is to be wired using single-core 70 °C thermoplastic-insulated cable with copper conductors. The design current for the single-phase circuit (I_b) is 17 A and ambient temperature is 25 °C. Protection against short circuit only is by a type B circuit breaker to BS EN 60898.

As $I_n \geqslant I_b$ then a 20 A device is selected.

Using the information contained in Appendix 4 of BS 7671, the method of installation is B from Table 4A2.

C_a = 1.03 from Table 4B1 for 25 °C and thermoplastic cable

C_g = 0.6 from Table 4C1 given the number of circuits is 5 in total installed as method B

C_i = 1 as no thermal insulation exists

C_f = 1 as the device is *not* to BS 3036.

$$I_t \geqslant \frac{I_n}{C_g C_a C_i C_f}$$

Therefore:

$$I_t \geqslant \frac{17}{1.03 \times 0.6 \times 1 \times 1} = 27.5 \text{ A}$$

So, using the correct cable selection table for single-core 70 °C thermoplastic cable in Appendix 4 of BS 7671, we can select a conductor cross-sectional area (csa). Using Table 4D1A, and in particular column 4 as the circuit is single-phase and the installation method is B, the value of I_t suitable for the current calculated is 32 A, which indicates a 4 mm^2 conductor size.

The equation for protection against short circuit only (when overload protection is not required) is appropriate for motor circuits where the motor starter provides overload protection and for circuits supplying fixed loads. BS 7671 allows its use for any fixed loads, for example, water heaters.

However, it is usual to provide overload protection unless it is impracticable, as for motor circuits (see BS 7671, Regulation 552.1.2).

The omission of overload protection is also allowed where unexpected disconnection would cause danger, as stated in BS 7671, Regulation 433.3.3.

If overload protection is not provided then the **adiabatic** (without loss or transfer of heat) check of BS 7671, Regulation 434.5.2 is carried out to ensure the cable is suitably protected.

The value I can be taken as the value of fault current that causes rapid disconnection of a device; it can be found in Appendix 3 of BS 7671.

Using the example on page 312, the required information is:

k = 115 from Table 43.1 of BS 7671

S = 4 mm^2

I = 100 A for a 20 A type B circuit breaker (as Appendix 3 of BS 7671) for a disconnection time of 0.1 seconds

Therefore:

$$t = \frac{115^2 \times 4^2}{100^2} = 21.16 \text{ seconds}$$

As the device will disconnect within 0.1 seconds, and the cable can safely carry 100 A fault current for 21.16 seconds before its final limiting temperature is reached, the circuit is acceptable.

However, the equation can be rearranged to demonstrate its objective more clearly, as follows.

$$I^2 t = k^2 S^2$$

$I^2 t$ is proportional to the thermal energy let through by the protective device under fault conditions, and $k^2 S^2$ is the thermal capacity.

KEY POINT

There are many circumstances where the overcurrent device provides fault protection but not overload protection. The most common is when the load is fixed or when a motor starter provides overload protection. In these circumstances, the cable rating must exceed the load but not necessarily the overcurrent device rating.

Adiabatic

The adiabatic equation can be used in several ways. It can be used to determine the time taken for a given cable to exceed its final limiting temperature. Alternatively, it can be used to determine the minimum csa to be able to withstand a fault current for a given duration without exceeding the final limiting temperature.

Current-carrying capacity tables

Tabulated current-carrying capacity I_t

Appendix 4 of BS 7671 provides current-carrying capacities of cables in certain defined conditions. Each table specifies:

- the cable type
- the ambient temperature
- the conductor operating temperature
- the reference method of installation.

The correct table for any particular cable is found by reference to Table 4A3 of BS 7671.

The tabulated cable current rating I_t is the current that will increase the temperature of the conductor of the cable from the tabulated ambient temperature, usually 30 °C, to the tabulated maximum conductor operating temperature (for example, 70 °C for thermoplastic cables and 90 °C for thermosetting) under the defined conditions.

For example, for a 4 mm² single-core 70 °C thermosetting cable, at an ambient temperature of 30 °C, enclosed in conduit in an insulating wall, we can see from Table 4D1A of BS 7671 that columns 2 and 3 are for method A (enclosed in a thermally insulated wall).

Therefore, I_t = 26 A for two cables single-phase a.c. or d.c. and 24 A for three or four cables three-phase a.c. or d.c.

Each table has a corresponding voltage drop table, with the exception of Table 4D5 of BS 7671, where the voltage drop is incorporated in the current-carrying capacity table.

Rating factors

If the actual conditions of installation do not meet the reference conditions, then rating factors are applied to the tabulated rating I_t as multipliers to obtain the cable's current rating (I_z) in the actual installation conditions.

$$I_z = I_t C_a C_g C_i C_f$$

where:

I_z is the current-carrying capacity of a cable for continuous service under the particular installation conditions concerned

I_t is the tabulated current-carrying capacity of a cable found in Appendix F of the IET On-Site Guide and Appendix 4 of BS 7671

C_a is the rating factor for the ambient temperature (see Table 4B of BS 7671 or Table F1 of the IET On-Site Guide)

C_g is the rating factor for grouping (see Table 4C of BS 7671 or Table F3 of the IET On-Site Guide)

C_i is the rating factor for conductors surrounded by thermal insulation

As long as the calculated value of I_z exceeds the value of I_n then the circuit is acceptable.

As we have seen before, I_t is the current-carrying capacity for specific cables, as given in BS 7671, Table 4D1A onwards. This method may be used where voltage drop is a greater consideration. For cables selected by voltage drop first, this formula will be used to verify that the selected cables are suitable for the intended current capacity. This method requires practice and experience but can prove to be more efficient, with fewer calculations needed.

KEY POINT

Where a formula has no mathematical operation symbols showing, it means the values should be multiplied.

ACTIVITY

What is the minimum csa of aluminium conductor that may be used?

Example:

Confirm the suitability for a 6 mm^2 70 °C thermoplastic, flat profile twin and cpc cable to be installed, enclosed in conduit within a thermally insulated wall. The ambient temperature is 25 °C and the conduit contains one other circuit. The circuit is protected by a 20 A type B circuit breaker to BS EN 60898.

Using the information contained in Appendix 4 of BS 7671, the method of installation is A from Table 4A2.

$C_a = 1.03$ from Table 4B1 for 25 °C and thermoplastic cable

$C_g = 0.8$ from Table 4C1, given the number of circuits is 2 in total installed as Method A

$C_i = 1$ as no further thermal insulation rating factor is required

$C_f = 1$ as the device is *not* to BS 3036

$I_t = 32$ A from Table 4D5 for flat profile cable

As we are confirming the cables suitability, we could use:

$$I_z = I_t C_a C_g C_i C_f$$

Therefore:

$$I_z = 32 \times 1.03 \times 0.8 \times 1 \times 1 = 26.4 \text{ A}$$

As the rating of the protective device (I_n) is 20 A and the resulting I_z is greater than this, the cable is suitably protected.

Cable selection for voltage drop

As well as selecting a cable for current-carrying capacity, a cable cross-sectional area (csa) must also be suitably sized to minimise voltage drop.

Voltage drop limit

BS 7671 requires that, under normal service conditions, the voltage at the terminals of any fixed current-using equipment shall be greater than that of the product standard.

Furthermore, in the absence of any product standard, the voltage drop should not exceed that for the proper working of the equipment.

As a result of this, the following guidance is given in Appendix 4 of BS 7671:

Voltage drop in consumers' installations

The voltage drop between the origin of an installation and any load point should not be greater than the values in the table below expressed with respect to the value of the nominal voltage of the installation.

Type of installation	Lighting	Other uses
Low-voltage installations supplied directly from a public low-voltage distribution system	3%	5%
Low-voltage installation supplied from private LV supply	6%	8%

Extract from Table 4Ab in Appendix 4 of BS 7671:2008 Requirements for Electrical Installations (the IET Wiring Regulations 17th edition) (2011)

Although experienced designers may make use of different tolerances to the above values, compliance with the table above will satisfy most installations.

Basic voltage-drop calculation

Single-phase circuits

In order to calculate the voltage drop in a single-phase circuit, the following information is required regarding the circuit:

- the value of milli-volts per ampere per metre (mV/A/m) voltage drop for the cable selected for current-carrying capacity (found in the voltage-drop table associated with the cable selection tables in Appendix 4 of BS 7671)
- the design current for the circuit (I_b) in amperes
- the total circuit length in metres.

ASSESSMENT GUIDANCE

To get a really accurate value where the loads are distributed along the cable, as in a lighting circuit, it would be necessary to calculate the voltage drop in each section.

Then we apply the equation:

$$\text{voltage drop (V)} = \frac{\text{mV/A/m} \times I_b \times L}{1000}$$

This will determine the voltage drop at the maximum operating temperature of the circuit.

To verify the voltage drop determined, the figure must be compared to the maximum permitted for the supply voltage (assumed to be 230 V) depending on the type of load.

For lighting circuits, the maximum would be:

$$230 \times \frac{3}{100} = 6.9 \text{ V}$$

For power circuits, the maximum would be:

$$230 \times \frac{5}{100} = 11.5 \text{ V}$$

Example:

A single-phase cooker is wired using 70 °C thermoplastic 6 mm^2 single-core cable in surface-mounted conduit (Method B), to a length of 10 m. The design current (I_b) is 29 A. Determine the suitability of the cable in terms of voltage drop.

From Table 4D1B (voltage-drop values) in Appendix 4 of BS 7671, the value of mV/A/m for a 6 mm^2 cable as column 3 (single-phase, method B) is 7.3 mV/A/m. Therefore:

$$\text{voltage drop} = \frac{7.3 \times 29 \times 10}{1000} = 2.1 \text{ V}$$

As the maximum permitted for a power circuit is 11.5 V, this is acceptable.

ACTIVITY

The supply voltage, nominally 230 V, will be divided between the load and the (L & N) conductors. The greater the voltage drop in the cables, the less voltage there is for the load. Less voltage; less current; less power.

A pure resistive load takes a current of 20 A at 230 V (4600 W). What would be the power output at 220 V assuming the load resistance remains the same?

Three-phase voltage drop

The voltage-drop tables in Appendix 4 of BS 7671 also include values of voltage drop in mV/A/m for three-phase circuits. These values assume that the loads are balanced, resulting in no neutral current. The permitted value of voltage drop now relates to the line-to-line voltage of 400 V (U) instead of the line-to-neutral value.

Voltage drop (per ampere per metre):   Conductor operating temperature: 70° C

Conductor cross-sectional area	Two-core cable d.c.	Two-core cable, single-phase a.c.			Three- or four-core cable, three-phase a.c.		
1	2	3			4		
(mm²)	(mV/A/m)	(mV/A/m)			(mV/A/m)		
1.5	29	29			25		
2.5	18	18			15		
4	11	11			9.5		
6	7.3	7.3			6.4		
10	4.4	4.4			3.8		
16	2.8	2.8			2.4		
		r	x	z	r	x	z
25	1.75	1.75	0.170	1.75	1.50	0.145	1.50
35	1.25	1.25	0.165	1.25	1.10	0.145	1.10
50	0.93	0.93	0.165	0.94	0.80	0.140	0.81
70	0.63	0.63	0.160	0.65	0.55	0.140	0.57
95	0.46	0.47	0.155	0.50	0.41	0.135	0.43

Extract from Table 4D4B of BS 7671:2008 Requirements for Electrical Installations (the IET Wiring Regulations 17th edition) (2011)

Example:

Verify the voltage drop for a three-phase 400 V motor with a design current (I_b) of 17.2 A. The circuit is wired using 4 mm² 70 °C thermoplastic insulated steel-wire armoured cable to a length of 24 m, clipped directly to a wall.

Using Table 4D1B (voltage-drop values) in Appendix 4 of BS 7671, for a 4 mm² steel-wire armoured cable, the value of voltage drop in mV/A/m, from column 4 (three or four-core cable, three-phase), is 9.5 mV/A/m. Therefore:

$$\frac{9.5 \times 17.2 \times 24}{1000} = 3.92 \text{ V}$$

The maximum permitted value of voltage drop permitted is:

$$400 \times \frac{5}{100} = 20 \text{ V}$$

Therefore the value of 3.92 is well within the permitted value.

Determining cable size by voltage drop

In some situations, voltage drop rather than current-carrying capacity ultimately affects the size of a conductor. This is common for circuits having long lengths but relatively small design currents. In this situation, the minimum value of mV/A/m can be determined by:

$$\text{max mV/A/m} = \frac{\text{max permitted voltage drop} \times 1000}{I_b \times L}$$

The value of mV/A/m is compared to the voltage-drop tables in Appendix 4 of BS 7671, in order to select a cable with a value lower than that determined. Following this, the cable may be proved for adequate capacity by selecting the particular cable-current capacity (I_t) and applying:

$$I_z = I_t C_a C_g C_i C_f$$

As long as the calculated value of I_z exceeds I_n, the circuit is satisfactory for both current capacity and voltage drop.

Correction for load power factor

For cables with conductors of cross-sectional area (csa) of 16 mm^2 or less, their inductances are not significant and only (mV/A/m)$_r$ values are tabulated.

For cables with conductors of cross-sectional area greater than 16 mm^2, the impedance values are given as (mV/A/m)$_z$ together with the resistive component (mV/A/m)$_r$ and the reactive component (mV/A/m)$_x$. This can be seen in Table 4D4B from BS 7671.

If the power factor of the load is not known, the (mV/A/m$_z$) value of voltage drop is used. This would include circuits such as distribution circuits, where the final circuits have multiple items in which power factor may affect the circuit but a fixed value is not known.

Where a more accurate assessment of voltage drop is required and the power factor (cos θ) is known, such as motor circuits, the following methods may be used.

For cables with conductors of cross-sectional area of 16 mm^2 or less, the design value voltage drop is determined approximately by multiplying the tabulated value of mV/A/m by the power factor of the load, cos θ:

$$\text{voltage drop} = \frac{\text{mV/A/m} \times \cos \theta \times I_b \times L}{1000}$$

For cables with conductors of cross-sectional area greater than 16 mm^2, the design value of mV/A/m is determined approximately as cos θ (tabulated (mV/A/m)$_r$) + sin θ (tabulated (mV/A/m)$_x$), and then this value of mV/A/m is used in the formula:

$$\frac{\text{mV/A/m} \times I_b \times L}{1000}$$

> **KEY POINT**
>
> The following information regarding load power factor is beyond the requirements of this course but is included here to provide further explanation.

> **KEY POINT**
>
> To determine sin θ from the given power factor (cos θ), the angle must be found. This is done on a calculator by using cos^{-1} θ to get an angle, then sin θ to determine the sin value.

Assessment criteria

3.6 Determine the size of conduit and trunking as appropriate to the size and number of cables to be installed

SmartScreen Unit 304
Handout 19 and Worksheet 19

KEY POINT

Remember that space factors refer to the amount of cable contained. So a 45 % space factor means 45 % cable, 55 % air.

Bend

The term 'bend' means a British Standard 90-degree bend. One double set is equivalent to one bend.

DETERMINING CONDUIT AND TRUNKING SIZES

Trunking and conduit systems

Trunking systems made of steel are commonly used in commercial and industrial premises. Conduit is used to take supplies to the various accessory outlet boxes. Plastic multi-compartment trunking is used in offices where different electrical services are required to be installed in close proximity to each other. Trunking and conduit, which is available in PVC or steel, provide good mechanical protection for cables and allow alteration or rewiring of the installation to be carried out easily. Training on threading and bending is required when steel systems are used.

There are, however, limits to the number of cables that can be installed in trunking and conduit, mainly in order to eliminate problems associated with the build-up of heat inside the trunking and conduit.

Spacing factors of wiring enclosures

Spacing factors are the free air allowances made around each cable installed in an a.c. conduit or trunking system. The method described below can be used to determine the size of conduit or trunking needed to accommodate multiple cables of the same or different sizes in accordance with BS 7671. It employs a unit system, with a factor allocated to each cable size. Tables in Appendix E of the IET On-Site Guide can be used to satisfy this. The sum of the factors for all the cables needed in the same enclosure is compared against the factors given for conduit, ducting or trunking.

Any **bends** in a conduit run have a bearing on the calculation, so a choice must be made between three different categories:

- For case 1 (conduit straight runs not exceeding 3 m in length), each cable and conduit size is represented by one single factor.

- For case 2 (conduit straight runs exceeding 3 m, runs of any length with bends or runs that are offset (sets) within the containment system), each conduit size has a variable factor, depending on the length of run and the number of bends or sets.

- For case 3 (trunking), each cable size and trunking size is allocated a factor.

A number of other factors also have a bearing on the number of cables that can be installed in conduit and trunking. These include:

- level of care during installation
- use of the space available

- tolerance in cable sizes
- tolerance in conduit and trunking.

The tables on the following pages can only give guidance on the maximum number of cables that should be drawn into an enclosure but should ensure an easy pull with low risk of damage to the cables. However, this method does not assess the electrical effects of grouping. It may also sometimes be more economical to divide the circuits concerned between a number of enclosures rather than have one very large enclosure.

As the grouping factor (C_g) is used to determine the cable cross-sectional area, the more cables installed, the larger the cross-sectional area required. This has a negative impact, so it is ultimately better to install fewer cables in the same enclosure.

KEY POINT

The maximum distance between draw-in points in conduit is 10 metres.

Example 1

▼ **Table E1**   Cable factors for use in conduit in short straight runs, taken from the IET On-Site Guide

Type of conductor	Conductor cross-sectional area (mm²)	Cable factor
Solid	1	22
	1.5	27
	2.5	39
Stranded	1.5	31
	2.5	43
	4	58
	6	88
	10	146
	16	202
	25	385

▼ **Table E2**   Conduit factors for use in short straight runs, taken from the IET On-Site Guide

Conduit diameter (mm)	Conduit factor
16	290
20	460
25	800
32	1400
38	1900
50	3500
63	5600

With reference to Table E1 of the IET On-Site Guide, a conduit run of 3 m with no bends and a total of ten 4 mm² thermoplastic stranded cables has a cable factor of $10 \times 58 = 580$.

Reference to Table E2 indicates that the minimum size for the conduit would be 25 mm diameter, which has a conduit factor of 800.

Example 2

▼ **Table E3** Cable factors for use in conduit in long straight runs over 3 m, or runs of any length incorporating bends, taken from the IET On-Site Guide

Type of conductor	Conductor cross-sectional area (mm²)	Cable factor
Solid or Stranded	1	16
	1.5	22
	2.5	30
	4	43
	6	58
	10	105
	16	145
	25	217

With reference to Table E3 of the IET On-Site Guide, a conduit run of 10 m with two bends and a total of ten 4 mm² thermoplastic stranded cables has a cable factor of 10 × 43 = 430.

Reference to Table E4 below indicates that the minimum size for the conduit would be 32 mm diameter, which has a conduit factor of 474.

▼ **Table E4** Conduit factors for runs incorporating bends and long straight runs, taken from the IET On-Site Guide

Length of run (m)	Straight				One Bend				Two Bends				Three Bends				Four Bends			
	16	20	25	32	16	20	25	32	16	20	25	32	16	20	25	32	16	20	25	32
1	Covered by Tables E1 and E2				188	303	543	947	177	286	514	900	158	256	463	818	130	213	388	692
1.5					182	294	528	923	167	270	487	857	143	233	422	750	111	182	333	600
2					177	286	514	900	158	256	463	818	130	213	388	692	97	159	292	529
2.5					171	278	500	878	150	244	442	783	120	196	358	643	86	141	260	474
3					167	270	487	857	143	233	422	750	111	182	333	600				
3.5	179	290	521	911	162	263	475	837	136	222	404	720	103	169	311	563				
4	177	286	514	900	158	256	463	818	130	213	388	692	97	159	292	529				
4.5	174	282	507	889	154	250	452	800	125	204	373	667	91	149	275	500				
5	171	278	500	878	150	244	442	783	120	196	358	643	86	141	260	474				
6	167	270	487	857	143	233	422	750	111	182	333	600								
7	162	263	475	837	136	222	404	720	103	169	311	563								
8	158	256	463	818	130	213	388	692	97	159	292	529								
9	154	250	452	800	125	204	373	667	91	149	275	500								
10	150	244	442	783	120	196	358	643	86	141	260	474								

Additional factors:
▶ For 38 mm diameter use 1.4 x (32 mm factor)
▶ For 50 mm diameter use 2.6 x (32 mm factor)
▶ For 63 mm diameter use 4.2 x (32 mm factor)

Example 3

▼ **Table E5**   Cable factors for trunking, taken from the IET On-Site Guide

Type of conductor	Conductor cross-sectional area (mm²)	PVC BS 6004 Cable factor	Thermosetting BS 7211 Cable factor
Solid	1.5	8.0	8.6
	2.5	11.9	11.9
Stranded	1.5	8.6	9.6
	2.5	12.6	13.9
	4	16.6	18.1
	6	21.2	22.9
	10	35.3	36.3
	16	47.8	50.3
	25	73.9	75.4

Notes:

1   These factors are for metal trunking and may be optimistic for plastic trunking, where the cross-sectional area available may be significantly reduced from the nominal by the thickness of the wall material.

2   The provision of spare space is advisable; however, any circuits added at a later date must take into account grouping, Regulation 523.5.

With reference to Table E5 of the IET On-Site Guide, a trunking run of 10 m with two bends and a total of 20 × 4 mm² and 10 × 16mm² thermoplastic stranded cables has a cable factor of (20 × 16.6) + (10 × 47.8) = 810.

Reference to Table E6 below indicates that the minimum size for the trunking would be 100 × 25 mm, which has a trunking factor of 993.

▼ **Table E6**   Factors for trunking

Dimensions of trunking (mm x mm)	Factor	Dimensions of trunking (mm x mm)	Factor
50 x 38	767	200 x 100	8572
50 x 50	1037	200 x 150	13001
75 x 25	738	200 x 200	17429
75 x 38	1146	225 x 38	3474
75 x 50	1555	225 x 50	4671
75 x 75	2371	225 x 75	7167
100 x 25	993	225 x 100	9662
100 x 38	1542	225 x 150	14652
100 x 50	2091	225 x 200	19643
100 x 75	3189	225 x 225	22138
100 x 100	4252	300 x 38	4648
150 x 38	2999	300 x 50	6251
150 x 50	3091	300 x 75	9590
150 x 75	4743	300 x 100	12929
150 x 100	6394	300 x 150	19607
150 x 150	9697	300 x 200	26285
200 x 38	3082	300 x 225	29624
200 x 50	4145	300 x 300	39428
200 x 75	6359		

Note: Space factor is 45% with trunking thickness taken into account.

Note that where other dimensions of cable or trunking other than specified in Tables E5 and E6 are used, an allowance of 45% space is needed.

The selection of protective conductors is covered in Chapter 54 of BS 7671.

Types of protective conductor

Earthing conductors and circuit protective conductors are both protective earthing conductors. This means they are intended to carry a current under earth fault conditions.

BS 7671 states that a protective conductor may consist of one or more of the following:

- a single-core cable
- a conductor in a cable
- an insulated or bare conductor in a common enclosure with insulated live conductors
- a fixed bare or insulated conductor
- a metal covering, for example, the sheath, screen or armouring of a cable
- a metal conduit, metallic cable management system or other enclosure or electrically continuous support system for conductors
- an extraneous conductive part complying with Regulation 543.2.6.

The adiabatic equation

Protective conductors carry very little current in normal operating circumstances (when there is no fault to earth). However, under earth fault conditions, they have to carry the fault currents for the duration of the fault. Fault currents may be very high but last for very short periods (until the overcurrent device operates). Consequently the csa of the protective conductor is not determined from the continuous ratings in Appendix 4 of BS 7671 but by the adiabatic equation of BS 7671 Regulation 543.1.3 (see below). This allows a much smaller cable to be used, hence the flat twin cables with a reduced protective conductor, commonly used in the UK in domestic and similar installations, and reduced cpc sizes when wiring single-core cable in conduit or trunking.

The adiabatic equation is:

$$S = \frac{\sqrt{I^2 t}}{k}$$

where:

S is the minimum cross-sectional area (csa) of the conductor in mm^2
I is the value in amperes (rms for a.c.) of fault current for a fault of negligible impedance that can flow through the associated protective device, with due account being taken of the current-limiting effect of the circuit impedances

t is the operating time of the disconnecting device in seconds, corresponding to the fault current I in amperes

k is a factor taking account of the resistivity, temperature coefficient and heat capacity of the conductor material, and the appropriate initial and final temperatures.

Values of k for protective conductors, depending on their type, material and method of installation, are as given in Tables 54.2 to 54.6 of BS 7671. Each table represents the different types of protective conductor or the different methods of installing it, as shown below.

Table 54.2	For protective conductors not bunched or incorporated in a cable. Examples include a separate earthing conductor at the intake or 'tails' position in a domestic installation or a separate 'overlay' cpc run with an armoured cable
Table 54.3	For protective conductors bunched or incorporated in a cable. This is the most common table as it covers all multicore composite cables, single-core cables in conduit and trunking
Table 54.4	For protective conductors that form sheaths or armours of a cable. Examples include the use of a steel-wire armouring as a cpc. However, this does not include MICC cable as a copper value is not given for this, so Table 54.2 is used
Table 54.5	For situations where a conduit or trunking is used as a common cpc. Although rare today, this is still an accepted method
Table 54.6	For bare conductors, such as metallic earth strapping, commonly used in main switch rooms (similar to that seen on lightning protection systems)

Values of k for protective conductors in various situations

Where the application of the formula produces a non-standard size, a conductor with the nearest larger standard cross-sectional area should be used.

Example:

Verify the suitability of the cpc for thermal constraints where a circuit has a total earth fault loop impedance of 2.3 Ω and is wired using 2.5/1.5 mm² 70 °C thermoplastic flat profile twin and cpc cable. A 20 A BS 88-3 device protects the circuit.

First we need to determine the earth fault current:

$$I_f = \frac{U_o}{Z_s}(A)$$

Therefore:

$$I_f = \frac{230\text{ V}}{2.3\text{ Ω}}(A) = 100\text{ A}$$

Using the graphs in Appendix 3 of BS 7671 (Figure 3A1), we can see that the device will disconnect in 0.7 seconds with a fault current of 100 A.

From Table 54.3 in BS 7671 (cpc incorporated in the cable), the value of k is 115.

The minimum size for thermal constraint (S) is determined using:

$$S = \frac{\sqrt{I^2 t}}{k}$$

Therefore:

$$S = \frac{\sqrt{100^2 \times 0.7}}{115} = 0.727\text{ mm}^2$$

In this particular situation, the minimum csa for the cpc required is 0.727 mm², which can be rounded to 1 mm², so the 1.5 mm² conductor is suitable under fault conditions.

The conductor csa can be calculated using the adiabatic equation or selected in accordance with Table 54.7 of BS 7671 shown below. This method is simpler than the use of the adiabatic equation but it may not give the most economical results.

Minimum cross-sectional area of protective conductor in relation to the cross-sectional area of associated line conductor (Source: Table 54.7 of BS 7671:2008 Requirements for Electrical Installations (the IET Wiring Regulations 17th edition), 2011)

Cross-sectional area of line conductor S	Minimum cross-sectional area of the corresponding protective conductor	Minimum cross-sectional area of the corresponding protective conductor
	If the protective conductor is of the same material as the line conductor	If the protective conductor is not of the same material as the line conductor
(mm^2)	(mm^2)	(mm^2)
$S \leqslant 16$	S	$\dfrac{k_1}{k_2} \times S$
$16 < S \leqslant 35$	16	$\dfrac{k_1}{k_2} \times 16$
$S > 35$	$\dfrac{S}{2}$	$\dfrac{k_1}{k_2} \times \dfrac{S}{2}$

where:

k_1 is the value of k for the line conductor, selected from Table 43.1 in Chapter 43 of BS 76761, according to the materials of both conductor and insulation

k_2 is the value of k for the protective conductor, selected from Tables 54.2 to 54.6 of BS 76761 (see above), as applicable.

According to Table 54.7 of BS 7671, the cpc needs to be the same size as the line conductor, for any circuit where the line conductor csa is up to 16 mm^2. This means that all twin and cpc cables with a reduced cpc size need to be verified by the adiabatic equation during the design process.

Where a circuit breaker is used as the protective device and high fault levels in the region of 3 kA and above are expected, the minimum protective conductor csa must be increased because of the increased energy let through by the circuit breakers. This can be seen on Table B7 of the IET On-Site Guide, which gives recommended minimum cpc cross-sectional areas relating to circuit breaker types and ratings.

Confirming cpc for earth fault loop impedance

As the total earth fault loop impedance directly affects the value of earth fault current, it is the responsibility of the designer to select the parts of the earth fault path, within the installation, to ensure disconnection times are met. This includes the values of resistance $R_1 + R_2$ for any distribution circuit and extends to the extremities of all final circuits.

Remember:

$$I_f = \frac{U_o}{Z_e + (R_1 + R_2)}$$

where:

I_f is the fault current

U_o is the nominal a.c. rms line voltage to earth

Z_e is that part of the earth fault loop impedance which is external to the installation

R_1 is the resistance of the line conductor of the circuit at conductor operating temperature

R_2 is the resistance of the circuit protective conductor of the circuit at conductor operating temperature.

Determining values of $R_1 + R_2$

$(R_1 + R_2)$ in ohms (Ω) is calculated by multiplying the resistances in milliohms per metre (mΩ/m) at 20 °C, from the table of values of resistance/metre below, by the length of the circuit L and a resistance correction factor. For the purpose of this publication, we will refer to this correction factor for temperature as C_r (see the table on page 330 for values of this factor). Also, to avoid confusion, we will refer to the values of resistance for a line and cpc in mΩ/m at 20 °C as r_1+r_2.

To calculate circuit (R_1+R_2), we apply the following formula:

$$(R_1 + R_2)\,\Omega = \frac{LC_r(r_1 + r_2)}{1000}\,\Omega$$

where:

C_r is a factor to correct resistances at 20 °C to the conductor maximum operating temperature (see the table on page 330 for factor C_r)

L is the circuit cable length

$(r_1 +r_2)$ is the resistance of the line conductor plus the protective conductor, in mΩ/m at 20 °C.

ACTIVITY

Calculate the resistance of 50 m of 6 mm² line and 1.5 mm² cpc at 20 °C.

Values of resistance/metre for copper and aluminium conductors at 20 °C in milliohms/metre (r_1+r_2) (Table I1 of the On-Site Guide)

Cross-sectional area mm²		Resistance (r_1+r_2) mΩ/m
Line conductor	Protective conductor	Copper
1.0	–	18.1
1.0	1.0	36.2
1.5	–	12.1

Cross-sectional area mm^2		Resistance (r_1+r_2) mΩ/m
Line conductor	Protective conductor	Copper
1.5	1.0	30.2
1.5	1.5	24.2
2.5	–	7.41
2.5	1.0	25.51
2.5	1.5	19.51
2.5	2.5	14.82
4.0	–	4.61
4.0	1.5	16.71
4.0	2.5	12.02
4.0	4.0	9.22
6.0	–	3.08
6.0	2.5	10.49
6.0	4.0	7.69
6.0	6.0	6.16
10.0	–	1.83
10.0	4.0	6.44
10.0	6.0	4.91
10.0	10.0	3.66
16.0	–	1.15
16.0	6.0	4.23
16.0	10.0	2.98
16.0	16.0	2.30
25.0	–	0.727
25.0	10.0	2.557
25.0	16.0	1.877
25.0	25.0	1.454
35.0	–	0.524
35.0	16.0	1.674
35.0	25.0	1.251
35.0	35.0	1.048

ASSESSMENT GUIDANCE

Composite cables such as twin and cpc tend to come with specific live-to-cpc cable ratios. Single cables may be installed in a mix that complies with the adiabatic equation.

| Cross-sectional area mm² | | Resistance (r_1+r_2) mΩ/m |
Line conductor	Protective conductor	Copper
50.0	–	0.387
50.0	25.0	1.114
50.0	35.0	0.911
50.0	50.0	0.774

Factor C_r to be applied to table of values of resistance/metre (pages 328–330) in order to calculate conductor resistance at maximum operating temperature for standard devices (Source: IET On-Site Guide)

| Conductor installation | Conductor insulation | | |
	70 °C thermoplastic (pvc)	85 °C thermoplastic (rubber)	90 °C thermosetting
Not incorporated in a cable and not bunched (Note 1)	1.04	1.04	1.04
Incorporated in a cable or bunched (Note 2)	1.20	1.26	1.28

Note 1: See Table 54.2 of BS 7671:2008 Requirements for Electrical Installations (the IET Wiring Regulations 17th edition), which applies where the protective conductor is not incorporated or bunched with cables and for bare protective conductors in contact with cable covering.

Note 2: See Table 54.3 of BS 7671, which applies where the protective conductor is a core in a cable or is bunched with cables.

Essentially, the correction factors in the above table raise the resistance of the conductor by 2% for every 5 Celsius degrees change in temperature.

Example:

Determine the total earth fault loop impedance (Z_s) for a radial final circuit, protected by a 20 A type B circuit breaker of length 25 m, wired in 70 °C thermoplastic insulated and sheathed cable, with 2.5 mm² line conductors and 1.5 mm² protective conductor. The external loop impedance Z_e is 0.35 Ω.

Using the cable resistance table, we can see that a 2.5/1.5 mm² combination gives a r_1+r_2 value of 19.51 mΩ/m at 20 °C and the correction factor is 1.20, as the cpc is incorporated in a sheath with the line conductor. So, using:

$$(R_1 + R_2) = \frac{L \times C_r \times (r_1 + r_2)}{1000}\Omega$$

We can determine:

$$(R_1 + R_2) = \frac{25 \times 1.20 \times 19.51}{1000} = 0.585 \ \Omega \text{ at } 20 \ °C$$

The fault current I_f is calculated using:

$$I_f = \frac{U_o}{Z_e + (R_1 + R_2)}$$

Therefore:

$$I_f = \frac{230}{0.35 + (0.585)} = 246 \text{ A}$$

When I_f is known, reference to the device characteristic will give the disconnection time. (See the type B circuit breaker characteristics from Appendix 3 of BS 7671 below.)

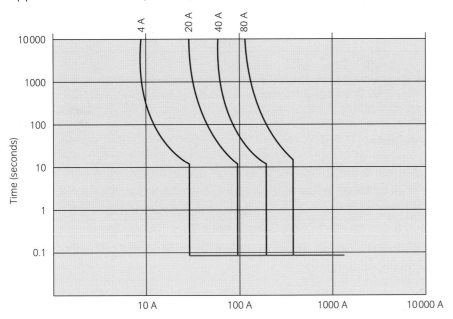

Type B circuit breaker time–current characteristic (Source: Fig 3A4 of Appendix 3 to BS 7671:2008 (2011) Requirements for Electrical Installations (the IET Wiring Regulations 17th edition))

In the example above, with a 20 A type B circuit breaker with a fault current I_f of 245A, the circuit breaker will operate in 0.1 seconds. This is well within the requirement of 0.4 seconds of Table 41.1 of BS 7671 (see table of maximum disconnection times for TN and TT systems on page 277).

Maximum earth fault loop impedances

For TN systems, maximum values of earth fault loop impedance are given in Table 41.2, 41.3 and 41.4 of BS 77671:2008 Requirements for Electrical Installations, depending on the type of protective device used. These tables are used where the nominal voltage to earth (U_o) is 230 V, and it removes the need to calculate the earth fault current.

As long as the value of earth fault loop impedance (Z_s) determined is within the values in the table, disconnection will be achieved.

This maximum fault loop impedance (Z_s) to achieve disconnection in the required time in the tables has been determined from:

$$Z_s \leqslant \frac{230}{I_a}$$

Hence, for the type B circuit breakers above, we can calculate the Z_s values, as shown in this table.

Rated current I_n	Fault current for 0.1 disconnection I_a	$Z_s = \dfrac{230}{I_a}$
6	30	7.670
20	100	2.300
40	200	1.150
80	400	0.575

ACTIVITY

Using the current characteristic tables in Appendix 3 of BS 7671, determine the maximum permitted values of Z_s for each of the following circuit breakers. Once determined, compare these values with Table 41.3 of BS 7671:

- 32 A type C
- 100 A type B
- 32 A type D.

KEY POINT

Remember: types of circuit breaker are selected to suit a particular load, not to suit earth fault loop impedance. If the earth fault loop impedance is too high, changing the rating or type of circuit breaker is not a suitable option.

ASSESSMENT GUIDANCE

Time–current curves are available from BS 7671 Appendix 3 or from fuse and switchgear manufacturers.

Example:

Determine the suitability of a circuit protected by a 6 A type B circuit breaker, to be wired in 70 °C thermoplastic flat profile twin with cpc cable, with a line and neutral conductor csa of 1.5 mm^2 and a cpc csa of 1.0 mm^2. The length (L) is 100 m. The supply and installation form a TN-C-S system with external impedance (Z_e) of 0.35 Ω.

First we consider:

$$Z_e + (R_1 + R_2) \leqslant Z_{s\,max}$$

So:

from Table 41.3 of BS 7671 or the table above, $Z_{s\,max}$ is 7.67 Ω

from the table on pages 328–330 (also Table I1 of the On-Site Guide and Table 13.4a of BS 7671), $(r_1 + r_2) = 30.2$ mΩ/m at 20 °C

from the table on page 330 (Table I3 of the On-Site Guide and Table 13.4b of BS 7671), C_r is 1.20 as the cpc is incorporated in the cable, and:

$$(R_1 + R_2) = \frac{L \times C_r \times (r_1 + r_2)}{1000}\ \Omega$$

Therefore:

$$(R_1 + R_2) = \frac{100 \times 1.20 \times 30.2}{1000} = 3.62\ \Omega$$

As:

$$Z_s = Z_e + (R_1 + R_2)$$

then:

$$0.35 + 3.62 = 3.97\ \Omega$$

As $Z_{s\,max}$ is 7.67 Ω this is satisfactory.

Where RCDs are used to meet disconnection times

In some cases, earth fault loop impedance values may be too high to satisfy disconnection times of standard protective devices. A solution to this is to install an RCD or RCBO.

As these devices disconnect at much lower fault currents, much higher earth fault loop impedance values are permitted.

Table 41.5 in BS 7671 gives maximum earth fault loop impedance values where earth fault protection is provided by an RCD. There is no maximum residual current rating for an RCD providing automatic disconnection of supply (ADS), but, in most installations, devices with residual current ratings ($I_{\Delta n}$) of 30 mA are commonly installed to provide additional protection.

When a circuit relies on an RCD as fault protection, it must be verified that the circuit still has suitable short circuit and, where required, overload protection.

KEY POINT

Additional protection should not be confused with ADS. Additional protection is required where there is greater risk of electric shock through the failure of basic protection, such as insulation, leading to direct contact with live parts, whereas ADS provides fault current protection.

ACTIVITY

Look in the index of BS 7671. List a sample of the locations where additional protection is required. From this list, work out the common risk and the reason for additional protection.

The knowledge of how wiring systems are integrated into a building is underestimated and often overlooked. Correct selection and use of equipment ensures that installations function correctly throughout their design life.

DIFFERENT TYPES OF CIRCUIT

BS 7671 Requirements for Electrical Installations (IET Wiring Regulations) state:

> The number and type of circuits required for lighting, heating, power, control, signaling, communication and information technology, etc shall be determined from knowledge of:
>
> (i)   location of points of power demand
>
> (ii) loads to be expected on the various circuits
>
> (iii) daily and yearly variations in demand
>
> (iv) any special conditions, such as Harmonics
>
> (v)   requirements for control, signalling, communication and information technology, etc
>
> (vi) anticipated future demand if specified.

This requirement demonstrates the importance of dividing an electrical installation into different circuit types to meet the different functions of a complete working electrical installation. The whole nature of electrical sub-distribution and final distribution for commercial installations has changed in the last few years. There is a demand for more residual current device (RCD) protection of final circuits, more metering and often more control to meet energy-saving targets.

By dividing circuits into types based on location and expected loads, system reliability can be enhanced, cable kept to sensible sizes and appropriate protective devices used. The designer will need to assess the current demand, taking **diversity** into account.

ASSESSMENT GUIDANCE

There are many wiring systems that could be used for a domestic installation. It could be wired in mineral-insulated copper-clad cable (MICC) but would be very expensive and difficult to alter at a later date. Conduit could be used but, again, is inflexible. Despite having some disadvantages, thermoplastic flat insulated and sheathed cable is usually suitable in terms of flexibility and ease of installation.

Diversity

The concept of diversity is based on the idea that not all circuits will be working at 100% all the time. Because of this a designer of an installation can reduce the design current on certain circuits of that installation.

Circuits can be categorised in the following way:

- **Lighting circuits** are provided in all installations. They are low power circuits with an almost constant load when lights are operating. For more detail, see the section on lighting circuits below.

- **Power circuits** are provided for socket outlet circuits. They require larger protective devices due to the demand for current and, in most instances, require RCD protection due to the additional risk presented by the portable equipment connected to the socket outlets. For more detail, see the section on power circuits, on pages 338–339.

- **Alarm and emergency systems** require additional protective measures, which include separating them from other circuits so that they are not affected by any faults on those other circuits. Further protection by wiring the circuit in fireproof cabling, may be considered.

- **Data and control circuits** are normally wired in different types of cable to a low voltage electrical system. However the electrical supplies need more rigorous cabling and earthing arrangements to reduce interference between the different wiring systems and to meet electromagnetic compatibility (EMC) requirements.

ACTIVITY

Diversity is applied to an installation to take into account the fact that not all lights, sockets, etc are in use all the time. Look around your home and count how many lights, out of the total installed, are actually on at a given time one evening.

KEY POINT

BS 7671 refers to 'mutual detrimental influence', which is simply the effect that two or more circuits may have on each other.

Lighting circuits

Lighting circuits are generally rated at 6 A but can in some installations be rated at 10 A or 16 A. Consideration must be given to the type of lighting point used for higher rated circuits due to requirements within Chapter 55 of BS 7671.

Depending on the type of wiring system, lighting circuits may be wired:

- three-plate: for circuits wired in **composite** cable such as thermosetting insulated and sheathed flat profile cable, commonly found in domestic dwellings

- conduit method: for circuits wired in single-core cables within a suitable containment system.

Composite cables

Composite cables are multi-cored, where the cores are surrounded by a sheath providing mechanical protection.

Three-plate

Because composite cables limit the freedom to take a single conductor between one point and another, lighting circuits wired in composite cables require more joints within the circuit. These additional joints in the cable are called loop-ins and are usually in the form of a third live terminal within a light point such as a ceiling rose. As the rose contains a third terminal, the system gets the name three-plate. Although it is traditional to install the three-plate at the light point, this isn't necessarily the case in all situations. Some installers will install the three-plate at a switch, normally when the wiring in the ceiling is difficult to access such as those in multi-storey flats or flat roofs.

The diagram below shows both the connections and that a two-core with cpc cable is used to connect the various points. Other light points can be installed on the circuit by looping off the ceiling rose. The circuit can be modified to allow two-way switching.

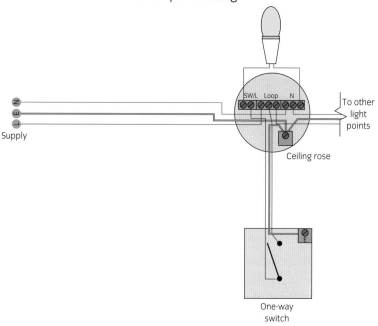

A three-plate one-way lighting circuit

Two-way switching is where one light point or luminaire can be controlled by two switches. By linking the two switches by a three-core and cpc cable, it allows the circuit to be opened or closed from either switch. Two-way switches do not actually open and close but divert the current from the common terminal (C) to either terminal L1 or L2. The two cables that link terminals L1 and L2 between the switches are called strappers.

To allow the light point or luminaire to be controlled from one or more further switches, the strappers need to be interrupted by one or more intermediate switches as shown on the next page.

A three-plate circuit controlled via two-way switches

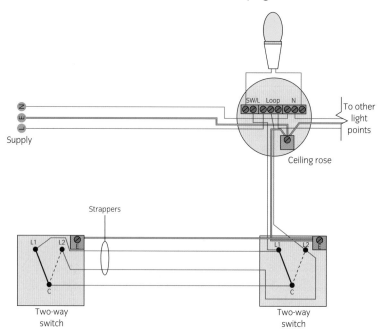

With three-plate systems, it can be seen from the diagrams that cable cores need jointing to enable them to be extended to a further point. One example of this is the common connection between the two way lighting that requires jointing in the intermediate switch.

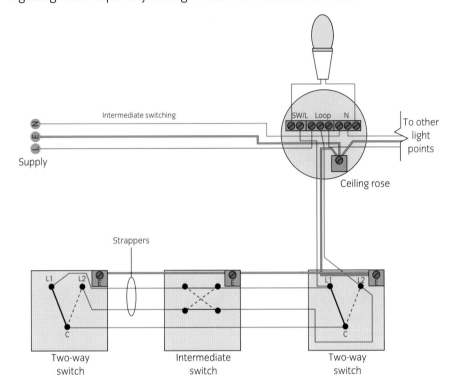

Three-plate with intermediate switching

Lighting systems wired in single-core cables can reduce the amount of jointing in cables as cables can be wired to the point where they are required.

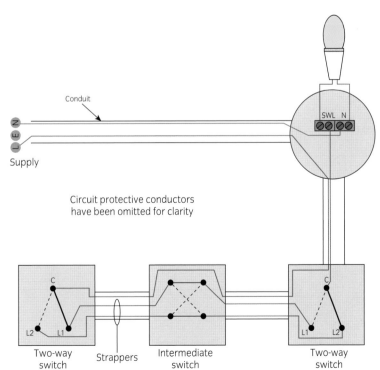

Two-plate intermediate circuit using single-core cables

Power circuits

Power circuits generally supply socket outlets but may also supply individual appliances. Power circuits may be wired in two different ways: ring-final and radial.

Ring-final circuits are the traditional means of wiring socket-outlet circuits within the UK. The reason for their use was to provide a high number of conveniently placed outlets adjacent to the loads. The circuit load is shared by two conductors in parallel. This also assists in improving voltage drop as the conductors being in parallel reduce the overall resistance.

Ring-final circuits are able to supply an unlimited number of outlets serving a maximum floor area of 100 m^2. Permanent loads, such as immersion heaters, should not be connected to a ring-final circuit. Ring-final circuits supplying kitchens should be arranged to have the loads equally distributed around the circuit. The plugs and fused spur units are normally fitted with 3 A or 13 A fuses.

With technology reducing the consumption of appliances and equipment, the need for a ring-final circuit is being questioned as they do have several disadvantages. If a ring-final circuit becomes open circuit, this may not necessarily be detected by the user as power will still be distributed to all socket outlets. Circuit conductors may then become overloaded leading to the possibility of fire. Earth fault loop impedances and circuit resistance values between live conductors may also increase due to this, leading to increased disconnection times. Testing continuity of ring-final circuits is also more time-consuming than testing the continuity of radial circuits. Designers today are very likely to specify radial socket-outlet circuits with a reduced nominal rating of protective device due to these reasons. This also reduces the inconvenience of many socket outlets being lost due to a single circuit failure.

ASSESSMENT GUIDANCE

Both ring-final circuits and radial circuits have restrictions imposed on them regarding the number of spurs and floor areas served. Look in the IET On-Site Guide and list these restrictions.

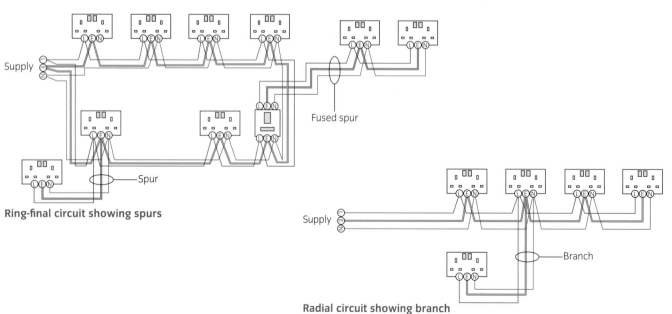

Supply

Fused spur

Spur

Ring-final circuit showing spurs

Supply

Branch

Radial circuit showing branch

Radial circuits may be selected to supply multiple socket-outlets for the reasons given on page 338 and they are also chosen as a means of supplying individual appliances or dedicated fixed appliance circuits.

SELECTING WIRING SYSTEMS, EQUIPMENT AND ENCLOSURES

Assessment criteria

4.1 State the criteria for correctly selecting wiring systems, equipment and enclosures as appropriate for systems

As was highlighted in Learning outcome 3, Regulation 5 of the Electricity at Work Regulations 1989 (EAWR) requires the duty holder (ie the designer in the case of selecting systems, equipment and enclosures) to ensure that all system components that are selected are appropriate to the intended use and rated accordingly.

The view of the Health and Safety Executive (HSE) is that, if electrotechnical work is undertaken in accordance with BS 7671 (the IET Wiring Regulations), it is likely to meet the requirements of the Electricity at Work Regulations 1989.

Assessment of general characteristics

You need to consider a variety of factors when selecting an electrical system.

An assessment must be made of the general characteristics of the installation, as detailed in the table below. The relevant chapter of BS 7671 is listed against each characteristic.

Characteristic	Chapter
The purpose for which the installation is intended to be used, its general structure and its supplies	31
The external influences to which it is to be exposed	32
The compatibility of its equipment	33
Its maintainability	34
Recognised safety services	35
Assessment for continuity of service	36

Purpose, structure and supplies

Within Part 3 of BS 7671, Chapter 31 requires a designer of an electrical installation to assess, during the design stage:

- maximum demand and diversity
- conductor arrangement and system earthing
- types of system earthing
- supplies
- division of installation
- external influences
- compatibility of characteristics and electromagnetic compatibility
- maintainability
- safety services
- continuity of service (need for back-up supplies).

While others may give the designer some of the above detail, such as supplies, it is still the designer's responsibility to assess suitability.

External influences

According to Regulation 512.2 of BS 7671, all equipment must be suitable for the situation in which it is to be installed and used, in terms of temperature, exposure to water, dust, mechanical impact and use, as well as other situations in which the equipment is to be used. These factors are listed in full in Appendix 5 of BS 7671, with each influence given a specific code. Designers must be aware of all the influences to which the installation is likely to be subjected.

Cable installation method

Like all equipment, the cabling system adopted for a location must be suitable for the environment. Table 4A2 of Appendix 4 of BS 7671 lists many cable installation methods. The common systems are briefly described in the table on the next page.

Table 4A2 reference number	Method	Description	Application
100–103	Flat twin with cpc	An insulated and sheathed cable with protective conductor. Installed in insulation	Very limited mechanical protection – plastic sheath. Generally used to install in walls, ceilings, floors, etc where building structure provides limited protection. Cheapest method
C	Steel wire armoured cable	Insulated and sheathed multicore cables with armouring	Clipped directly to walls
D			Installed underground
E			On cable tray
E	Non-armoured cable	Insulated and sheathed multicore cables without armouring, including fire-retardant cables	For use on cable tray
C	Armoured or non-armoured cables with sheath including fire-retardant cables		Clipped directly to a wood or masonry surface
B	Steel conduit (black enamelled) and trunking	Insulated single or multicore cables with or without separate protective conductor. Also includes multi-compartment trunking such as dado trunking	Good mechanical protection, for use indoors. Training on threading and bending required
B	Steel conduit (galvanised)	Insulated single or multicore cables with or without separate protective conductor	Good mechanical protection, for use outdoors and indoors. Training on threading and bending required
B	Surface-mounted PVC conduit and trunking	Single-core insulated cables with separate protective conductor or multicore cable. Also includes multi-compartment trunking such as dado trunking	Very suitable for outdoor and corrosive environments, eg agricultural. Cheap, easy to install
A	Conduit in a thermally insulated wall	Single-core insulated cables with separate protective conductor or multicore cable	Steel or PVC
B	Floor trunking	Insulated single or multicore cables with or without separate protective conductor	Provides accessible sockets in very large open-plan offices. Expensive
Not classified in table	Busbar trunking	Fitted with busbars	For machine shops and similar. Allows flexibility if machines are to be moved. Expensive

Common cable installation methods (Source: Table 4A2 of Appendix 4 of BS 7671:2008 Requirements for Electrical Installations (the IET Wiring Regulations 17th edition), 2011)

Twin and earth cable

Wiring systems for different environments

Different types of wiring system are available because not all systems are suitable for all environments or compatible with all types of construction.

Domestic installations

These installations tend to use twin and earth cables (referenced as 6242Y by wholesalers) or similar types of sheathed cable. This type of cable is ideal because it is compatible with the building products used and the finishes expected by the end user. It is also easy to install because it is flexible and can be bent and shaped into tight locations. Although other types of cables could be used, the relatively low cost of this type is also important in the domestic market, which is very cost sensitive.

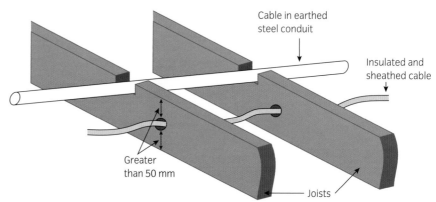

Cable in earthed steel conduit

Insulated and sheathed cable

Greater than 50 mm

Joists

Installation method when crossing joists

There are various types of containment systems suitable for PVC/PVC 6242Y wiring systems. However, these tend to be used in non-domestic circumstances (see pages 346–348).

Commercial installation

There are so many different types of commercial building that this category probably includes the widest range of electrical installations. Commercial installations can be found in large converted houses, listed buildings, purpose-built portal frame buildings, multi-storey offices and shopping complexes with pre-tensioned concrete structures. In all these scenarios the installations need to be compatible with the structure and the finished layouts need to be either as cost effective or as sympathetic to the aesthetics of the building as possible.

Any, or all, of the following systems may need to be considered in commercial installations.

Supply distribution systems are usually found in the 'back-of-house' areas or common structural cores of commercial buildings where special finishes to the building fabric are not required. In such functional areas it would not be unusual to see steel wire armoured (SWA) cable with low smoke and fume sheathing, which is usually described as XLPE SWA LSF cabling.

Fire alarm and similar systems need to be wired in fire-resistant cabling such as mineral-insulated, copper-clad cables (MICC), a soft-skinned fire-resistant cable such as FP200 or, where enhanced cabling is required by Building Regulations, FP200 plus or similar products.

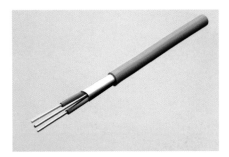

FP200 gold fire-resistant cable

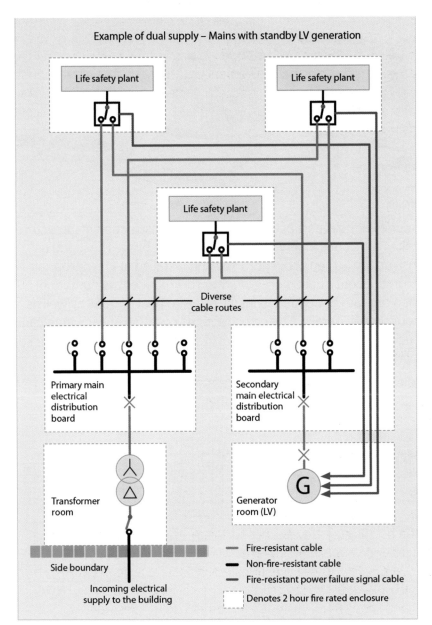

Typical fire-fighter lift arrangement using 2-hour fire-resistant cabling

Safety of life systems are fire resistant, usually for 120 minutes. They include FP600 ranges of cabling, which are tried and tested for fire-fighting equipment, sprinkler pumps, lifts, etc. These systems are usually separated from other systems so that they are not affected by faults on any of the other systems.

FP600S fire-resistant armoured cables

For systems in public areas, it is likely that the installation will be in recessed conduits and that single cables will be concealed in the structure. Office developments often have raised floors and power is provided through floor boxes plugged into underfloor 63 A busbar systems. With the rise in prefabrication of site services, modular wiring (often known as 'plug n play' systems) may even be used.

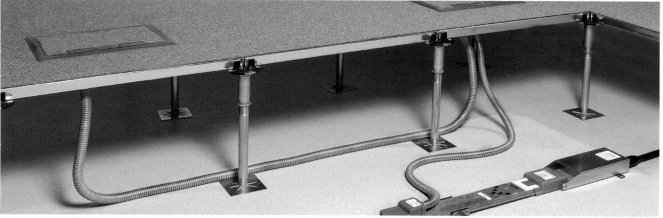

Underfloor busbar and plug-in floor box arrangement

Modular wiring is part of a larger modular construction manufactured off site. It includes pipework, ductwork and cabling. The modular construction is then brought to site and installed at an appropriate time in the construction programme. This reduces the number of skilled operatives needed to install the electrical system, as well as labour time generally. This often reduces costs. Modular wiring cannot be used for fire-fighting or safety of life systems as these must be kept separate from other systems.

ACTIVITY

Which section of BS 7671 specifically deals with swimming pools?

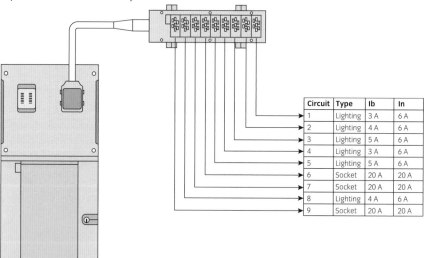

Circuit	Type	Ib	In
1	Lighting	3 A	6 A
2	Lighting	4 A	6 A
3	Lighting	5 A	6 A
4	Lighting	3 A	6 A
5	Lighting	5 A	6 A
6	Socket	20 A	20 A
7	Socket	20 A	20 A
8	Lighting	4 A	6 A
9	Socket	20 A	20 A

Modular wiring diagram

Agricultural installations

These installations require the wiring system to be resistant to the elements, livestock and vermin. As there is an increased risk of damage, armoured cables, conduit and sealed trunking systems are used. PVC conduit is more suitable as it will not corrode.

Industrial installations

These installations are often located in harsh environments. Large amounts of SWA cable are installed on ladder racking along main routes from switch rooms. Where individual cables drop to serve distribution boards or other equipment, they are usually supported in baskets or trays.

Local wiring from distribution boards is normally fed to the relevant outlet through thermoplastic singles in conduit and trunking systems or through thermoplastic SWA cabling on ladder and basket arrangements.

Single insulated cables

Wiring systems using single insulated cables are common in industrial and commercial environments. The insulation is generally thermoplastic or LSF XLPE (low smoke and fume cross-linked polyethylene cable). These types of wiring systems are vulnerable to mechanical damage, so they must be contained in either complete trunking or a conduit system.

Installations in hazardous environments

Hazardous environments require robust cabling systems to prevent any possibility of ignition or of triggering an explosion. The majority of large supply cables are SWA cables on ladder racking.

Ladder racking containing SWA cable

Steel wire armoured cable

Where fire resistance is required, mineral-insulated, copper-clad (MICC) cables are used, providing strength, integrity as well as fire resistance.

An oil refinery is obviously a hazardous environment with the risk of explosion

A petrol forecourt, although familiar, is still a hazardous environment due to the risk of ignition

Containment systems

Most containment systems are not exclusive to a particular type of wiring system or installation.

Cable baskets

Cable baskets can be used to contain many types of cable that have more than basic insulation, including PVC/PVC 6242Y cables, MICC, soft-skinned fireproof cables and, on a smaller scale, steel wire armoured cables. Cable baskets are often used for Category 5A and Category 6 TCP/IP data cabling.

IT cables on basket

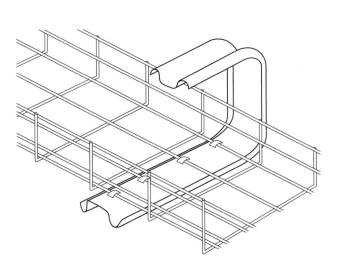

Methods of fixing baskets

Ladder racking and tray work

These containment systems are predominantly for industrial installations, although tray work can also be used for smaller armoured cables and MICC as well as soft-skinned cables. There are various types and grades of tray work to suit different cables and installations.

Returned flange traywork

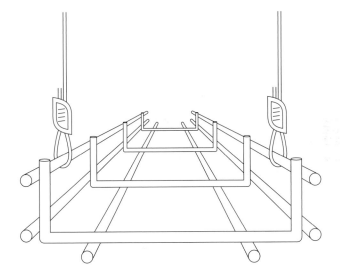

ASSESSMENT GUIDANCE

Trays can be bent, using a tray bender. Ladder racking is too heavy to be bent and manufactured fittings are used.

Ladder racking is used almost exclusively for armoured cables and typically where large amounts of heavy cables are installed, because it is strong enough to withstand the weight.

Ladder racking

SmartScreen Unit 304
Handout 19 and Worksheet 19

Conduit systems

In the UK, conduit systems are used in a number of ways. They include complete systems for single thermoplastic or XLPE cables with just basic insulation. This gives mechanical protection to prevent damage to the insulation and reduces the risks of shock, or fault hazard.

Conduit systems can be made of plastic, galvanised metal, stainless steel or black enamelled steel. Where there is risk of damp or corrosion, PVC or galvanised conduit should be selected. Where strong mechanical protection is required, galvanised conduit is normally used.

Galvanised conduit and fittings, courtesy of Lasnek

Black enamel finished metal conduit and coupling

Conduit systems are also used in part for protection of wiring system 'drops' to electrical equipment, etc so that walls or sections of building work can be completed and the cables pulled in later.

Couplings

Lengths of conduit are connected by couplings. Where a conduit enters a box that does not have a threaded entry, it is terminated by a coupling and a brass bush to complete the mechanical connection.

This stop end box has been converted into a back entry box using a conduit coupling and male bush arrangement. In proprietary assemblies, the box would have a manufactured thread into which a conduit with a male thread could be screwed

Systems of joining conduits, including back entry boxes, which can be manufactured or fabricated

Where two systems cannot be screwed together, a running coupling might be fabricated on site using two couplings, a nipple, a locknut and a length of conduit with an extra long thread. However running couplings create a potential weakness in the conduit and are best avoided.

Examples of conduit fittings: bends, through and outlet boxes, courtesy of Lasnek

Saddles

Spacer-bar saddles are the most commonly used method of securing conduit. Distance saddles are used where conduit is fixed to uneven or damp surfaces. Where hygiene is important, hospital saddles are used to allow easy cleaning of the shaped saddle and around the conduit.

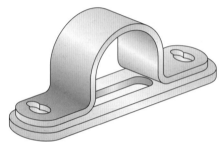

A spacer-bar saddle

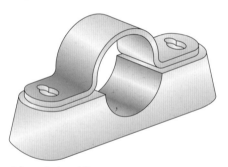

A hospital saddle

SmartScreen Unit 304
PowerPoint presentation 18,
Handout 18 and Worksheet 14

Trunking systems

In the UK, most of the wiring contained in trunking systems is single insulated cable. However, trunking systems can be used to carry many different types of cables, depending only on whether they will fit. The systems are available in a range of shapes with different numbers and profiles of compartments. Standard earth fault loop impedance values are often less on site than calculated. This is because conduit and trunking systems, and their support systems, can work as parallel conductors in real life, but cannot realistically be calculated.

KEY POINT

Mini-trunking is just a small version of larger PVC trunking. It can be supplied with a self-adhesive backing, to allow it to be fixed easily to a wall.

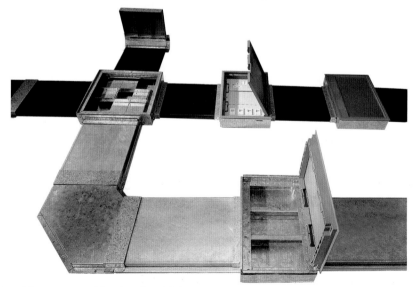

A trunking system with a bend and jointing section

Dado trunking

Dado trunking systems are often found in offices where there are large numbers of data and power outlets. They provide a multi-compartment system allowing data, telephone and power cables to be kept separate and distributed around the work space. They also offer flexibility in that the positions of outlets can easily be changed, to a degree, any time after installation to meet changing requirements.

Dado trunking in cross section

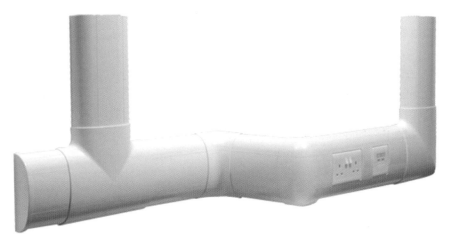

Dado trunking profile

Fixings

The choice of fixings depends very much on the combined weight of the containment system and the cables in it. Often, many kilograms of cable need to be supported. For example, larger cable such as four-core 240 mm^2 steel wire armoured cable can weigh in the region of 13 kg/m. So a simple fixing with a round headed wood screw and wall plug is often not enough.

Simple wood screws and plugs are often not strong enough to fix containment systems and the cables in them

Anchor bolt fixing

Anchor brackets

It is quite likely that specialist advice from proprietary fixings manufacturers will be required, along with confirmation from the structural engineer that the building structure can take the load, for example, whether the floor above can cope with additional loading from below. Where the load is considerable, substantial fixings such as anchor bolts or anchor brackets will be required.

WHAT YOU NOW KNOW/CAN DO

Learning outcome	Assessment criteria	Page number
1 Understand the characteristics and applications of consumer supply systems	*The learner can:*	
	1 Explain the characteristics and applications of consumer supply systems	254
	2 Specify the arrangements for electrical installations and systems with regard to provision for: ☐ isolation and switching ☐ overcurrent protection ☐ earth fault protection.	259
2 Understand the principles of internal and external earthing arrangements for electrical installations for buildings, structures and the environment	*The learner can:*	
	1 Explain the key principles relating to earthing and bonding	270
	2 Explain the key principles relating to the protection of electrical systems	275
	3 Explain the operating principles, applications and limitations of protective devices	279
	4 Specify what is meant by the terms relating to earthing and the function of earth protection: ☐ earth fault loop impedance ☐ protective multiple earthing (PME).	282
3 Understand the principles for selecting cables and circuit protection devices	*The learner can:*	
	1 Explain how external influences can affect the choice of wiring systems and enclosures	285
	2 State the current ratings for different circuit protection devices	289
	3 Specify and apply the procedure for selecting appropriate overcurrent protection devices	289
	4 State what is meant by diversity factors and explain how a circuit's maximum demand is established after diversity factors are applied	293
	5 Specify and apply the procedure for selecting a suitably sized cable	301
	6 Determine the size of conduit and trunking as appropriate to the size and number of cables to be installed.	320

4 Understand the principles and procedures for selecting wiring systems, equipment and enclosures	*The learner can:*	
	1 State the criteria for correctly selecting wiring systems, equipment and enclosures as appropriate for systems.	339

ASSESSMENT GUIDANCE

The assessment for this unit is in three sections.

- Task A: written project/assignment
- Task B: written short-answer paper

Task A (written project assignment)

- Task A is based on a scenario with plans.
- You will be expected to carry out planning activities or answer short questions.
- The assignment is open book using all available resources such as the internet.

Task B (written short-answer paper)

Task B consists of questions that require short answers of one or two short sentences. The paper consists of 30 questions and 2 hours, 30 minutes are allowed.

- For any exam, always arrive well before it starts.
- You may take BS 7671 and the IET On-Site Guide and a non-programmable calculator into the examination.
- Written exams should be completed in black or blue ink unless colours are specifically required for cable/terminal identification.
- Plan your progress through the exam. Do not spend so much time on one question that you leave insufficient time to finish answering others.
- If you need clarification, ask the invigilator, who will not supply technical answers to the questions.

Before the assessment

- You will find some questions in the section below to test your knowledge.
- Make sure you go over these questions in your own time.
- Spend time on revision in the run-up to the assessment.

OUTCOME KNOWLEDGE CHECK

1   Which supply arrangement relies on the consumer's earth electrode for the connection to the general mass of earth?

2   At which point of an installation does the consumer's installation start?

3   Name two circuits or situations that require emergency switching.

4   Identify a device that combines overload protection and additional protection against electric shock.

5   State what is meant by fusing factor.

6   State the principal of operation of an RCD.

7   What is meant by discrimination as applied to an installation?

8   What is meant by diversity as applied to an electrical installation?

9   State a wiring system suitable for a fire alarm system.

10   State why it is necessary to have additional earth electrodes on a TN-C-S supply system.

11   What is the method of installation, from BS 7671, for a surface metallic conduit?

12   Determine, using the appropriate method in the IET On-Site Guide, the design current (I_b) for a 12 kW 230 V electric cooker where the cooker control has an integral socket outlet.

So $\frac{12000}{230}$ = 52.18 A – 10 A = 42.18 × 0.3 (for 30%) = 12.66 + 10 A + 5 A (for socket) = 27.66 A

13   Determine, using BS 7671, the instantaneous tripping current for each of the following circuit breakers.

a)   32 A type B

b)   16 A type C

c)   32 A type D

14   List three external influences that must be considered when installing equipment outdoors.

15   State the table number and column number in BS 7671 used to select the maximum permitted current carrying capacity (I_t) for a 2.5 mm^2 single-core 70°C thermoplastic cable intended for a single-phase circuit installed as method B.

GLOSSARY

A

Abrade To scrape or wear away.

Adiabatic The adiabatic equation can be used in several ways. It can be used to determine the time taken for a given cable to exceed its final limiting temperature. Alternatively, it can be used to determine the minimum csa to be able to withstand a fault current for a given duration without exceeding the final limiting temperature.

Audit To conduct a systematic review to make sure standards and management systems are being followed.

Authorising engineer An engineer, usually chartered, appointed to check the knowledge and experience of those operating or working on specialist systems, to make sure that all the employer's legal duties are met.

Azimuth Ideally the modules should face due south, but any direction between
Azimuth refers to the angle that the panel direction diverges from facing due south

B

Bend The term 'bend' means a British Standard 90-degree bend. One double set is equivalent to one bend.

Bill of quantity A contract document comprising a list of materials required for the works and their estimated quantities.

Bioaccumulate Bioaccumulation occurs when substances such as pesticides or heavy metals gradually build up in the body of an organism, such as a human or other animal. These substances are not flushed through the body and a damaging amount can collect over time.

Blackwater Water that is contaminated by sewage or chemicals. Blackwater generally comes from kitchen sinks and toilets.

Breaking capacity This is the maximum value of current that a protective device can safely interrupt.

Business opportunity In this context, the opportunity to make profit from the work or contract.

C

CENELEC CENELEC (Comité Européen de Normalisation Électrotechnique) is the European Committee for Electrotechnical Standardization.

Chartered engineer An engineer with professional competencies through training and experience, also registered with the Engineering Council, which is the British regulatory body for engineers. The title chartered engineer (C Eng) is protected by civil law.

Civil law Law that deals with disputes between individuals and/or organisations, in which liability is decided and compensation is awarded to the victim.

Colour rendition (sometimes called rendering) The ability of the light emitted by a lamp to show objects in their true colour. For example, some LED lamps have a blue-ish light which can make some objects appear blue.

Combustible Able to catch fire and burn easily.

Competent person Recognised term for someone with the necessary skills, knowledge and experience to manage health and safety in the workplace.

Compliance The act of carrying out a command or requirement.

Composite cables Composite cables are multi-cored, where the cores are surrounded by a sheath providing mechanical protection.

Consignment A batch of goods.

Contamination The introduction of a harmful substance to an area.

Contract administrator The person named in the contract with the contractual power to change matters or items of work that will cause a contract variation.

Critical path The sequence of key events and activities that determines the minimum time needed for a process such as building a construction project.

D

Diversity Diversity is where the designer of an installation can reduce the design current of an installation on certain circuits. This is because it is known that not all circuits will be working at 100% all the time.

Dry-lined wall An inner wall that is finished with plaster-board rather than a traditional wet plaster mix.

Duty holder The person in control of the danger is the duty holder. This person must be competent by formal training and experience and with sufficient knowledge to avoid danger. The level of competence will differ for different items of work.

E

Earth fault loop impedance The impedance of the earth fault current loop starting and ending at the point of earth fault. This impedance is denoted by the symbol Zs.

Efficacy The measure of light output of a lamp compared to the energy used by the lamp measured in lumens per watt.

Effluvia Emissions of gas or odorous fumes given off by decaying waste.

Egress Leaving or exiting the site.

Enabling Act An enabling Act allows the Secretary of State to make further laws (regulations) without the need to pass another Act of Parliament.

Environment The environment is the land, water and air around us.

Exposed conductive parts Conductive parts of equipment, such as metal casings, that can be touched and are not normally live but can become live when basic insulation fails.

Extraneous conductive parts Conductive parts (eg metal, gas and water pipes) liable to introduce a potential (generally earth potential) and not forming part of the exposed metallic parts of the building's structure. These are then connected to the main earthing terminal, which in turn forms a circuit to the general mass of earth.

F

Fatality Death.

Hazard A hazard is usually defined as anything with the potential to cause harm (eg chemicals, a fault on electrical equipment, working at height).

Hazardous substance Something that can cause ill health to people.

Hot work Work that involves actual or potential sources of ignition and done in an area where there is a risk of fire or explosion (eg welding, flame cutting, grinding).

I

IMD Insulation monitoring device. An end-of-line resistor together with a fire detection system is an example.

Impedance Impedance (symbol Z) is a measure of the opposition that a piece of electrical equipment or cable makes to the flow of electric current when a voltage is applied. Impedance is a term used for alternating current circuits. The elements of impedance are:
(The effects of impedance on current such as leading and lagging are covered in Unit 302, see page 154.)

Incorporated engineer A specialist (also called an engineering technologist) who implements existing technology within a particular field of engineering, entitled to use the title IEng.

L

Liability A debt or other legal obligation in order to compensate for harm.

Load current The amount of current an item of equipment or a circuit draws under full load conditions.

M

Manual handling The movement of items by lifting, lowering, carrying, pushing or pulling by human effort alone.

N

Near miss Any incident that could, but does not, result in an accident.

P

PEN Protective earthed neutral. The PEN conductor is both a live conductor and a combined earthing conductor.

Personal protective equipment (PPE) All equipment, including clothing for weather protection, worn or held by a person at work, which protects that person from risks to health and safety.

PME Protective multiple earthing. This term is commonly used to describe a TN-C-S system.

Practical completion The point at which a construction project is virtually finished, when the last percentage of monies are paid by the client, the responsibility for insuring the construction transfers to the client and the architect or contract administrator signs the certificate of practical completion. At this point there will still be a few minor snags to sort out (eg scratches in paintwork) or insignificant items to be completed.

Q

Quantify Estimate guideline costs.

R

RCM Residual current monitor. This is much like an RCD but does not disconnect the circuit. Instead it gives a warning that a fault is present.

Risk The chance (large or small) of harm actually being done when things go wrong (eg risk of electric shock from faulty equipment).

S

Soakaway A system where rainwater or surface water is collected, then slowly discharged into the ground. The soakaway system would normally be below topsoil level to avoid waterlogging.

Statute A law made by Parliament as an Act of Parliament.

System This covers all and any electrical equipment which is, or may be, connected to an electrical energy source, and includes that source.

T

Tender An offer to carry out work, supply goods or buy land, shares or any form of asset for a fixed price, usually in competition with others.

Toolbox talk A method of communicating safety issues (as a supplement to an induction) or of giving continuous training on particular techniques or methods.

Toxic Poisonous.

W

Water bowser A transportable water tank or tanker.

Wholesome Term used in law to indicate water that is suitable for drinking.

Work package A 'collection' of work associated with one product on a project, eg all processes associated with the design, manufacture and installation of the windows or of the sprinkler installation.

ANSWERS TO ACTIVITIES AND KNOWLEDGE CHECKS

Answers to activities and knowledge checks are given below. Where answers are not given it is because they reflect individual learner responses.

UNIT 301 UNDERSTANDING HEALTH AND SAFETY LEGISLATION, PRACTICES AND PROCEDURES

Activity answers

Page

3   An induction session.

6   Ears muffs/defenders, gloves.

8   Safety officer.

11   Make sure the load mass is within your capabilities; warn people of your actions; make sure the passageway is clear of obstructions.

12   Such as Ribbmaster.

13   Securely replace the floorboard and/or surround it with barriers and notices. If this is beyond your level of responsibility, report it immediately.

16   By the use of a grinding wheel.

17   No trailing lead to trip over; generally lighter than mains drills; reduced risk of shock.

19   For example, blue or yellow Artic flex; Butyl heat resistant flex.

19   Yellow, blue, red.

20   No exposed conductors; terminations correctly tightened; conductors terminated to fill most of terminal; conductors correctly identified.

21   c) A continuous check should be carried out and records kept.

25   Should be treated by a trained first aider and taken to A&E to check on Tetanus protection and stitching if necessary. Accident report to be completed.

Page

27   Send them to a local hospital for examination. It is always a good idea if someone takes them.

31   No.

37   Ladder may slip; tiredness standing on rungs; falling tools, etc.

39   Gloves are commonly worn to protect against cuts and scratches. Cutting metal conduit or trunking would be a common application.

42   Any from, but not exclusively: running round blind corners, jumping out and scaring somebody, making a loud sudden noise, throwing objects, etc.

43   Before and after use.

45   Solvent adhesives; cutting pastes

46   Split rungs or treads, broken/ frayed ropes, cracks painted over.

48   IPAF card.

50   Any four not shown on page 50.

54   Smoke alarms/detectors. Mercury switches.

55   To wholesaler/manufacturer for disposal.

56   Radioactive material.

57   When protected by 30 mA RCD.

57   Friction between two materials can generate static, eg when scuffing your feet on a carpet with man made fibres or taking off a nylon boiler suit whilst wearing a wool jumper.

60   Route is clear of obstacles.

63 Move it/report to supervisor.

65 Any height above floor level.

66 Return to store for disposal.

69 Oil floats on water, fire will spread.

70 Clear area, report to supervisor.

Outcome knowledge check answers

Pages 75–78

1 c)	2 c)	3 b)	4 d)	5 b)
6 c)	7 c)	8 b)	9 b)	10 a)
11 c)	12 d)	13 c)	14 a)	15 b)
16 b)	17 c)	18 b)	19 c)	20 b)

UNIT 302 UNDERSTANDING ENVIRONMENTAL LEGISLATION, WORKING PRACTICES AND THE PRINCIPLES OF ENVIRONMENTAL TECHNOLOGY SYSTEMS

Activity answers

Page

81 Imperial conduit, PVC cables, switchgear and control gear, accessories, luminaires.

84 For example, televisions, DVD players, computers, laptops, tablets, mobile phones and white goods and small kitchen appliances.

86 Cleaning paint brushes.

88 Lead – old underground cables, steel – conduit or trunking, brass – plugs and sockets, contacts, copper – cables, aluminium – conductors, PVC – cable insulation, rubber – cable sheathing, cardboard – packaging.

90 45 sockets × 2 × 0.2 m = 18 m; 5 rings × 2 tails × 0.3 m =3 m; total waste = 21 m.

95 Examples may include fruit, DVDs, meat from supermarkets, etc.

101 Circulating pump.

105 Due South.

105 The wind over the roof can cause the panel to act like an aircraft wing, producing a force which can pull the panel off the roof.

110 Both about 90% (British Gas, 2013).

112 Rock or similar impervious material.

Page

114 Accumulator tank (used to store excess hot water).

116 CO_2 is a greenhouse gas which contributes to global warming.

118 a) Fibre glass or rockwool, b) foam, c) double glazing, d) polyethylene pipe lagging.

120 Biomass is quick growing plant material that takes in carbon dioxide when growing and releases it when burnt, so the overall effect is zero. When coal is burnt, carbon that has been stored for millions of years is released.

123 For example, access via narrow country lanes, narrow or weak bridges, sufficient storage in the event of road blockage due to snow or floods etc, provision of alternative heating system.

125 A gas that, when present in the atmosphere, traps heat emitted from the Earth.

128 There is a risk of electrocution from PV systems as they generate electricity in daylight conditions. There may be an output even in very dull conditions.

130 a) A roofing specialist or qualified PV installer, b) a qualified electrician.

131 To change the direct current produced by the PV array to alternating current.

135 All test results are required.

138 To light his holiday home.

141 Birds.

142 Feed-in tariff.

143 Migratory fish such as salmon and sea trout.

148 527.2 Sealing of wiring system penetrations. In particular, 527.2.1 – where wiring systems pass through the building structure, the openings must be sealed to the degree of fire protection required.

149 Rodents such as rats and mice; insects such as wasps and hornets.

152 Heat engine.

155 Non-statutory.

157 Water butts (barrels).

158 Bore holes, reservoirs, rivers etc.

161 No.

165 It would be necessary to have two separate waste water systems – one discharging waste from toilets, etc and the other for the greywater.

Outcome knowledge check answers

Page 173–174

1	c)	2	d)	3	a)	4	b)	5	b)
6	c)	7	d)	8	b)	9	b)	10	b)
11	a)	12	c)	13	a)	14	b)	15	b)
16	d)	17	b)	18	c)	19	b)	20	d)

UNIT 303 UNDERSTANDING THE PRACTICES AND PROCEDURES FOR OVERSEEING AND ORGANISING THE WORK ENVIRONMENT

Activity answers

Page

181 Typically – height 178 mm; width 101 mm; depth 115 mm (including window).

RCCB: BS EN 60898, terminal capacity 35 mm^2, 32 A.

182 Example: Electrical Contractors' Association.

186 Layout or location drawing/plan.

190 Weather, delays in material supplies, labour disputes, etc.

193 Various systems are available. Could be fixed floodlights at 110 V/230 V or festoon-type lighting.

194 4–6 weeks for simple works.

195 Collected, compacted/flattened and sent for recycling.

Page

197 See HSE construction information sheet no. 50.

198 63.5 V.

203 A quantity surveyor may check that the materials used on a particular job are the ones actually specified by the client.

204 They have the specific knowledge required for ever-changing modern systems.

208 A suitable list of trades will have no 'impossibilities' such as roofers arriving before bricklayers.

209 Part P in particular.

212 As the dispute is about you and is between your manager and the college, it would be inappropriate to involve anybody else.

220 From the list in Appendix 2 (page 48), regulations that are relevant are 4 and 15-12 in EAWR.

220 See Regulation 411.8 in BS 7671.

223 Make sure all obstacles/obstructions are moved out of the way and that everyone is warned of the activity.

224 The centre of gravity will be the point of balance. The load should be steadied as it is transported.

225 Dust mask, safety boots, hand cream/gloves.

227 Ignorance can quickly lead to an accident or damage to equipment.

228 An unforeseen incident causing physical damage or injury.

230 Refer to your supervisor.

232 The Electricity at Work Regulations, 1989.

235 Yes, it is a requirement for all those on the NVQ scheme.

235 Able to certify own work.

244 A delivery timed so that it arrives just before it is needed on site. This saves on storage and guards against possible pilferage and damage.

Outcome knowledge check answers

Page 252

1 Specification, layout plans, manufacturers' catalogues, internet, etc.

2 By use of signing in and signing out book; by supplying name badges; use of security staff, etc.

3 Polystyrene: specialist disposal; cardboard: package up for recyling.

4 Small offcuts of cabling to be sent for recycling. Longer lengths for store and reuse. Usable electrical accessories back to storage for use elsewhere.

5 Remove as much of the furniture, etc as possible. Use floor coverings and dust sheets. Fit dust-producing machinery with filters or vacuum suction.

6 a) Your supervisor.

b) Variation order.

7 Any three from: description of work carried out, address/site, materials used, time taken.

8 a) A-B-D-E-F (37).

b) Critical path becomes A-C-E-F (39) so job extended by 2 days.

9 a) That all items are present by checking against the original order.

b) That no items are damaged.

10 With the aid of the technical leaflet and a simple diagram, explain the operation and need for quarterly testing to ensure that the mechanism still works.

11 Arriving on site on time, being dressed in the correct clothing and PPE, being polite at all times.

12 A circuit diagram shows the relationship between the connections.

A wiring diagram shows the relationship between the components.

13 Schedule of inspections, schedule of test results.

14 Take-off sheet.

15 a) Records detail of extra work that has been requested. Must be signed by the client/representative before work commences.

b) Records day-to-day operations, deliveries, staff absences, etc.

c) Records notifiable accidents on site.

Activity answers

Page

259 a) Main switch; b) double-pole switch;
c) withdraw plug from socket;
d) circuit breaker / fuse.

260 Check that the device you selected is suitable for the following:

a) Isolation – cuts off all required poles and lockable

b) Mechanical maintenance – suitable load switching capability

c) Emergency switching – latching push button type or similar

d) Functional switching – suitable rating for full load switching and user friendly.

262 $16 \times 1.5 = 24$ A.

265 The circuit breaker cannot be held closed against a fault.

266 BS 7671 states that the residual current rating must not exceed 30 mA.

267 Green.

271 Protective multiple earthing.

274 Cannot rely on water pipe, which may have been partly replaced by plastic pipe.

276 Drill, jigsaw, angle grinder, chaser, vacuum cleaner, etc.

278 The following are approximate answers from the graphs: Type B = 20 s; Type C = 32 s; Type D = 38 s.

280 63 A.

282 16 A Type C – 160 A, 40 A Type B – 200 A; yes.
40 A Type B – 200 A, 63 A Type C – 630 A; yes.
50 A Type D – 1000 A, 100 A Type C – 500 A; no. Taken from Appendix 3, 0.4 to 5 s disconnection currents.

Page

283 Suitable drawing showing fault, earth electrode, return path through ground, transformer electrode, winding and line supply.

285 Ambient temperature, external heat sources, presence of water, presence of solid foreign bodies, presence of corrosive or polluting substance, impact, vibration and others as listed in Regulation 522.

290 a) £4.62; b) £9.24; c) £2.40.

291 You will still come across them in existing installations.

296 For 100 W lamp, $\frac{100}{230} = 0.43$ A

For 18 W lamp, $\frac{18}{230} = 0.08$ A

299

$I_{b \text{ load}}$	I_n BS EN 60898	I_t cable pvc/pvc
15	16	16
25	32	37
3	6	16
60	63	64
41	45	47

311 $I_{max} = 7000/230 = 30.43$ A

First 10 A plus 30% of remainder:

10 A + 30% of 20.43 = 16.1 A + 5 A for socket = 21.1 A

$I_n = 25$ A

Correction factors: Ca = 25 °C = 0.94;

$I_z = \frac{25}{0.94} = 26.6$ A, 2.5 mm^2 = 27 A

Volt drop = I_b × length × v_d/am

$$\frac{21.1 \times 15 \times 18 \text{ mv}}{1000} = 5.7 \text{ V}$$

315 16 mm².

317 $R = 230/20 = 11.5\ \Omega$, $P = \dfrac{V^2}{R} = \dfrac{220^2}{11.5} = 4208$ W

or $\left(\dfrac{220}{230}\right)^2 \times 4600 = 4208$ W

324 By placing thermometers along the cable run.

328 50 × (3.08 + 12.1) = 0.759 Ω.

332 Figures the same for 0.4 s and 5 s: 32 A Type C = 0.72 Ω×, 100 A Type B = 0.46 Ω, 32 A Type D = 0.36 Ω.

333 As per Regulation 411.3.3. Areas of increased risk of electric shock, external installations, bathrooms, swimming pools, etc. The common risks are high humidity, contact with earth or lack of clothing. This increases the risk of shock as body resistance is reduced.

342 So that standard floorboard fixing nails do not penetrate the cables.

344 Section 702.

345 They are likely to be damaged / twisted by the addition of extra conduit.

346 MICC is commonly used in underground ducts to feed the pumps.

348 20 mm and 25 mm.

349 16 mm, 20 mm, 25 mm and 32 mm. Larger sizes to special order.

Outcome knowledge check answers

Page 354

1   TT system.

2   At the outgoing meter terminals or supplier's DP switch if fitted.

3   Motor circuits, HV discharge lighting, petrol filling stations or suitable alternatives .

4   Residual current breaker with overload (RCBO).

5   The ratio of the fuse rating to the fusing current.

6   Under normal conditions the line and neutral currents are equal and the flux on each coil in the RCD cancels out. When an earth leakage flows, an out of balance occurs in the coils, inducing a current in the trip coil. This causes the RCD to trip.

7   Discrimination in an installation exists when it is designed so that the protective device closest to a fault operates first.

8   The ratio expressed as a percentage of the actual connected load to the maximum possible connected load.

9   MICC or FP/firetuf.

10 If the PEN conductor becomes broken, the consumer's metalwork could become live. The extra electrodes provide an alternative return path back to the supply transformer start point.

11 From table 4A2, method B.

12 Design current is found by: first 10 A is rated at 100 %, then 30 % of the remainder; allow 5 A for a socket outlet.

13 a) 160 A.   b) 160 A.   c) 640 A.

14 Three from:

- ingress of water
- mechanical damage
- temperature ranges
- direct sunlight (solar radiation)
- corrosion.

15 Table 4D1A, column 4.

INDEX

energy efficient lighting 93–4
energy rating system 95
enforcement, HSW Act 8–9
engine burners 152–3
England, building regulations 98–100
environment
    definition 80, 356
    impact reduction 90, 93–5
    legislation 79–89
    waste 80–95
Environment Act 1995 82–3, 223
Environment Agency 149
environmental factors 14, 42, 63–4, 243, 340–1
environmental hazard reporting 92
environmental impacts 86–7, 140, 148
environmental legislation 79–89
Environmental Protection Act 1990 (EPA) 81–2
Environmental Site Audits (ESA) 149
environmental suitability of products 179
environmental technology systems 79, 96–176
    biomass 120–6
    building regulations 97–100
    co-generation 96, 151–6
    electricity-producing 96, 127–50
    greywater reuse systems 164–70
    heat pumps 108–20
        air source 116–20
        ground source 111–16
    heat-producing 96, 101–26
    micro-hydro-electric 143–50
    micro-wind 136–43
    planning 97
    rainwater harvesting 159–64
    solar thermal systems 101–8
    water conservation 96, 157–70
environmentally friendly materials and products 93–5

EPA see Environmental Protection Act 1990
Equality Act 2010 223
equipment
    compatibility 198–9
    environment 243
    for outdoor use 59
    hazardous malfunctions 65
    inspection failures 20
    product specifications 179
    PUWER 6, 18, 24, 44, 65, 220
    regulations 18, 220
    selection principles 339–41
    structure 244
    suitability 243–4
    users 244
    visual inspection 19–20
erection of access equipment 48–9
ESA see Environmental Site Audits
ESQC see Electrical Safety, Quality and Continuity Regulations 2002
estimation 206, 242–3
European Union (EU) 8, 51–3, 255
evacuated-tube collectors 102–3
evaluation of hazards 55–6
evaporators 109
excess electricity 127
exempt appliances 123–4
explosion risks 52, 345–6
explosions 29–30, 58
explosive substances 52
exposed conductive parts 356
external earthing arrangements 270–84
external influences 285–6, 340
external lighting circuits 58
extra-low voltage systems 276
extraneous conductive parts 356
eye protection 38

F

F2508 record 28
falls 12–14, 44, 62–4
farms 274, 345

fatalities 28, 356
fault currents 261, 278–9, 289
    adiabatic equation 324–7
    impedance 289, 327–33
    temporary systems 58
faults 43–4
    circuit discrimination 280–2
    diagnosis and rectification 23–4
    earthing arrangements 271–5
    maximum disconnection times 277–8
    portable electrical equipment 19–20
    protective provisions 275–6
    shock hazards 56–7
    terminations and connections 20–1
feed-in tariffs 156
fibres 54
filtering
    greywater 166–8
    rainwater 160
final circuits 296–301
fire alarms 69, 343
fire extinguishers 69
fire-fighter lifts 343
fire-resistant cable 343, 345–6
fires 57, 68–9, 338, 343, 345–6
    procedures 29–30, 69
first-aid 40–1
fish 147, 149
fittings 350
five steps to risk assessment 35
fixings 351
flammable substances 12, 52
flat thermoplastic cables 288
flat thermosetting cables 288
flat-plate collectors 102
floor warming installations 300
flooring 12–13, 62
flow, hydro-electric systems 144, 147
flue gases 122–3
flues 154
fluid category 5 risk 165
fluorescent tubes 83, 84, 88–9
foam fire extinguishers 69